CHAPMAN & HALL/CRC
Monographs and Surveys in
Pure and Applied Mathematics   **144**

HIGHER ORDER

DERIVATIVES

CHAPMAN & HALL/CRC
Monographs and Surveys in
Pure and Applied Mathematics **144**

HIGHER ORDER

DERIVATIVES

Satya N. Mukhopadhyay

in collaboration with P. S. Bullen

CRC Press
Taylor & Francis Group
Boca Raton London New York

CRC Press is an imprint of the
Taylor & Francis Group, an **informa** business

A CHAPMAN & HALL BOOK

CRC Press
Taylor & Francis Group
6000 Broken Sound Parkway NW, Suite 300
Boca Raton, FL 33487-2742

First issued in paperback 2019

ISBN-13: 978-1-4398-8047-0 (hbk)
ISBN-13: 978-0-367-38174-5 (pbk)

Library of Congress Cataloging-in-Publication Data

Mukhopadhyay, Satya N.
   Higher order derivatives / S.N. Mukhopadhyay.
     p. cm. -- (Monographs and surveys in pure and applied mathematics)
   Includes bibliographical references and index.
   ISBN 978-1-4398-8047-0 (hardcover : alk. paper)
   1. Derivatives (Mathematics) 2. Differential calculus. I. Title.

QA325.M85 2012
515'.33--dc23   2011042819

This book is dedicated to

My Family

Contents

Preface

The concept of the first-order ordinary derivative is as old as the invention of the calculus. Since then the concept of the first-order ordinary derivative has been extended to higher order derivatives in various directions. The nth-order ordinary derivative $f^{(n)}$ of a function f is the first order derivative of its $(n-1)$st-order derivative $f^{(n-1)}$ and has no special importance. But, the higher order derivatives other than the ordinary one are particularly interesting in general because they are derivatives for which the nth order derivative can exist without the $(n-1)$st-order derivative existing. For instance, the classical Riemann derivative is an example of this type and plays an important role in the theory of trigonometric series.

Higher order derivatives of different types have been considered by several authors with the results appearing in various journals over a long period of time. We consider these higher order derivatives and study the relations between them. It is hoped that the resulting monograph will be particularly helpful to those young mathematicians who wish to pursue their studies in this branch of real analysis.

We cannot claim to have considered all known types of higher order derivatives and suggestions for further inclusion and improvements in general are welcome.

Introduction

The concept of higher order derivatives is useful in many branches of mathematics and its applications. Among the higher order derivatives, the Peano derivative is the most well known. This notion started from the viewpoint of approximating a function by polynomials and its origin is in the paper of G. Peano [139]. This derivative was subsequently considered, under various different names, by A. Genocchi, [68] and C.J. de la Vallée Poussin [174], and later by J.C. Burkill [28], A. Denjoy [45], J. Marcinkiewicz & A. Zygmund [106], J. Marcinkiewicz [105], A. Zygmund [194], and E. Corominas [37]. Finally in 1954, H.W. Oliver [134] gave the concept the name "Peano derivative" and made a systematic study of the Peano derivative. After this C.E. Weil [184] began to work on this derivative and since then extensive work has been done by Weil and many other authors [3,10,13–15,17,19,23,24,26,27,33,41,52,57,59,60,63,65,70,71,73,74,82,93–95, 97,100–102,124,127,128,135,137,138,142,151,159,161,181,183,185,195,196].

The Riemann* derivative is considered by A. Denjoy [45] and E. Corominas [37] under the name generalized derivative. They proved that this derivative exists finitely if and only if the Peano derivative exists finitely, and then the values are equal. The infinite case is considered in [26]; in this paper, there are some lacunæ pointed out in [94], which are filled up in [197]. A.M. Russel [145] gave this derivative its present name, Riemann* derivative, and used it to study functions of bounded variations of higher order. This derivative has been found to be useful as well for the study of functions of higher-order absolute continuity [43,114].

The de la Vallée Poussin derivative, also called the symmetric Peano derivative, was introduced under the name generalized derivative by C.J. de la Vallée Poussin himself [174]; he used it to study various properties of Fourier series. Since then, this derivative has been studied and used by many authors [7,9,16,20,22,25,30,34,36,38,47,72,77,78,84,87–92,105,108–110,115, 116,119,121–123,128,144,146,147,155,159,162,165–168,172,173,193,194,196].

The Cesàro derivative originated from the Cesàro summability of series. Introduced by J.C. Burkill to define the Cesàro–Perron integral [28,29], these derivatives were studied in detail by W.L.C. Sargent [150–152]. In addition, there is the work of J.A. Bergin [10], who studied the Cesàro derivatives proving, in certain cases, their equivalence to the Peano derivatives (see also [99]).

J.C. Burkill later introduced symmetric Cesàro derivatives [30] to define the symmetric Cesàro–Perron integral, which is useful in solving the so-called coefficient problem in the theory of trigonometric series; see also [25]. Higher order symmetric Cesàro derivatives were introduced in [22] to define the SC_nP-integral (see also [38]).

The concept of the Borel derivative was introduced by E. Borel, calling it the average derivative (derivée moyenne.) First-order Borel derivatives, both unsymmetric and symmetric, were studied by A. Khintchine [86] and by J. Marcinkiewicz & A. Zygmund [106]. Then W.L.C. Sargent [149] made an intensive study of the first-order unsymmetric Borel derivative (see also C.J. Neugebauer [133]). The symmetric Borel derivative was used by J. Marcinkiewicz & A. Zygmund [106] and by P.S. Bullen & S.N. Mukhopadhyay [25] to consider a trigonometric integral.

The L_p-derivative originated from the work of A.P. Calderón & A. Zygmund [32] and was considered in [159]. This derivative appears in [2, 5, 50, 58, 133, 136, 189, 196]. The symmetric L_p-derivative is considered in [2, 52, 131, 190].

The Abel derivative has a very special nature. Its origin can be found in a dormant state in the theory of trigonometric series and occurred in the work of A. Rajchman and A. Zygmund when they considered the Riemann and Abel summability of trigonometric series (see [196; p. 353, Lemma 7.6]). The definition of the Abel derivative is not given there, but is defined by S.J. Taylor [163]. Taylor only defined the second-order derivative and used it to introduce his Abel–Perron integral, which is helpful in solving the coefficient problem for Abel summable trigonometric series. As in the case of the de la Vallée Poussin derivative, the existence of the second-order Abel derivative of a function at a point does not imply the existence of the first-order Abel derivative at that point. To help with this, the concept of Abel smoothness is defined. The seeds of this and of higher-order Abel derivatives are in [122] and [121], respectively, and they are fully defined in the text.

The Laplace derivative is introduced in [160, 161]; its theory is still being developed [198]. Symmetric Laplace derivatives are introduced in the text in a very natural way.

The symmetric Riemann derivative was found to be convenient for studying the Riemann summability of trigonometric series. Both unsymmetric and symmetric Riemann derivatives of higher order are considered and studied by J. Marcinkiewicz [105] and J. Marcinkiewicz & A. Zygmund [106]. Symmetric Riemann derivatives of higher order are the special study of P.L. Butzer & W. Kozakiewicz [31] and T.K. Dutta & S.N. Mukhopadhyay [48]. Further work on both unsymmetric and symmetric Riemann derivatives of higher order can be found in [71, 73, 74, 81–83, 125, 148, 175–180, 182, 196]. J. Marcinkiewicz & A. Zygmund [106] defined another derivative of higher order that is of interest, which we call the MZ-derivative.

J.M. Ash [2], following a suggestion of A. Zygmund, introduces a system, a finite set of real numbers satisfying certain conditions, and uses this system to define a generalized derivative of higher order. Suitable specializations of the system lead to the symmetric and unsymmetric Riemann derivatives and to the MZ-derivative. This work is continued in [4,6,64,75,76,126]. The derivative of Ash is such that, if any of the derivatives of Peano or de la Vallée Poussin exists finitely, then the appropriate derivative of Ash also exists with the same value. This equivalence does not extend to the case of derivates, however, although a suitable modification of the Ash definition can be made to allow for this. That is, the modified Ash definition not only includes the Peano and de la Vallée Poussin derivatives, but their corresponding derivates as well.

As has been mentioned, higher order derivatives occur in the study of convexity and bounded variation of higher order (see [1,16,21,35,39,40,69,70, 85,111–113,119,129,140,141,143,154,188,192] and [11,43,114,130,145,166], respectively).

In the case of symmetric derivatives, smoothness is sometimes helpful in various discussions (see above paragraph on the Abel derivative). Thus, if a function f has a de la Vallée Poussin derivative of order n at x, then while f has a derivative of order $n-2$ at x it may not have a derivative of order $n-1$ at x. Then smoothness of order n may compensate for the nonexistence of this derivative of order $n-1$, or if f is not smooth, then it may be quasi-smooth; see [1,18,46,49,51,55,56,80,104,132,153,156,158,169–171].

We have not considered the approximate analogues of the derivatives and smoothness discussed above, but papers on these topics are included in the bibliography. This bibliography is surely not complete and any suggestion for further inclusion is welcome.

Although much work has been done on the Peano and de la Vallée Poussin derivatives, there is a large amount of work to be done on the other higher order derivatives as their properties remain often virtually unexplored. The purpose of this book is to introduce to any newcomer interested in the field of higher order derivatives to the present state of knowledge. The background required is that of only basic advanced real analysis and, although the special Denjoy integral has been used, a knowledge of the Lebesgue integral should suffice.

The book contains two chapters. Chapter 1 contains 16 subsections where the various higher order derivatives are introduced. Chapter 2 contains 24 subsections where the relations between these derivatives are given.

Chapter 1

Higher Order Derivatives

1.1 Divided Differences of Order n

Let n be a fixed positive integer and let f be a real valued function defined on the set $\{x_0, \ldots, x_n\}$ of $n+1$ distinct points. The nth order difference is defined by

$$Q_n(f; x_0, \ldots, x_n) = \frac{Q_{n-1}(f; x_0, \ldots, x_{n-1}) - Q_{n-1}(f; x_1, \ldots, x_n)}{x_0 - x_n}, \quad n \geq 2,$$
(1.1.1)

with

$$Q_1(f; x_i, x_j) = \frac{f(x_i) - f(x_j)}{x_i - x_j}, \quad i \neq j, \ i, j = 0, \ldots, n.$$
(1.1.2)

A simple inductive argument, given below, shows that

$$Q_n(f; x_0, \ldots, x_n) = \sum_{i=0}^{n} \frac{f(x_i)}{\prod_{\substack{j=0 \\ i \neq j}}^{n} (x_i - x_j)} = \sum_{i=0}^{n} \frac{f(x_i)}{\omega'(x_i)},$$
(1.1.3)

where

$$\omega(x) = \prod_{j=0}^{n} (x - x_j).$$

☐   In fact, if $n = 1$, then (1.1.3) follows from (1.1.2). Suppose that (1.1.3) holds for $n = m$ and let $n = m+1$. Then, from (1.1.1), we have, since (1.1.3) holds for $n = m$,

$$(x_0 - x_{m+1})Q_{m+1}(f; x_0, \ldots, x_{m+1})$$
$$= Q_m(f; x_0, \ldots, x_m) - Q_m(f; x_1, \ldots, x_{m+1})$$
$$= \sum_{i=0}^{m} \frac{f(x_i)}{\prod_{\substack{j=0 \\ i \neq j}}^{m} (x_i - x_j)} - \sum_{i=1}^{m+1} \frac{f(x_i)}{\prod_{\substack{j=1 \\ i \neq j}}^{m+1} (x_i - x_j)}$$
$$= \frac{f(x_0)}{\prod_{j=1}^{m} (x_0 - x_j)} + \sum_{i=1}^{m} \frac{f(x_i)}{\prod_{\substack{j=0 \\ i \neq j}}^{m} (x_i - x_j)}$$
$$- \sum_{i=1}^{m} \frac{f(x_i)}{\prod_{\substack{j=1 \\ i \neq j}}^{m+1} (x_i - x_j)} - \frac{f(x_{m+1})}{\prod_{j=1}^{m} (x_{m+1} - x_j)}$$

$$= (x_0 - x_{m+1}) \frac{f(x_0)}{\prod_{j=1}^{m+1}(x_0 - x_j)} + \sum_{i=1}^{m} \frac{f(x_i)(x_i - x_{m+1} - x_i + x_0)}{\prod_{\substack{j=0 \\ i \neq j}}^{m+1}(x_i - x_j)}$$

$$- \frac{f(x_{m+1})(x_{m+1} - x_0)}{\prod_{j=0}^{m}(x_{m+1} - x_j)}$$

$$= (x_0 - x_{m+1}) \left[\frac{f(x_0)}{\prod_{j=1}^{m+1}(x_0 - x_j)} + \sum_{i=1}^{m} \frac{f(x_i)}{\prod_{\substack{j=0 \\ i \neq j}}^{m+1}(x_i - x_j)} + \frac{f(x_{m+1})}{\prod_{j=0}^{m}(x_{m+1} - x_j)} \right]$$

$$= (x_0 - x_{m+1}) \sum_{i=0}^{m+1} \frac{f(x_i)}{\prod_{\substack{j=0 \\ i \neq j}}^{m+1}(x_i - x_j)}$$

and so (1.1.3) is established for $n = m + 1$.   □

It follows, from (1.1.3), that for any two functions f and g and for any two real numbers α and β

$$Q_n(\alpha f + \beta g; x_0, \ldots, x_n) = \alpha Q_n(f; x_0, \ldots, x_n) + \beta Q_n(g; x_0, \ldots, x_n). \quad (1.1.4)$$

It follows, also from (1.1.3), that $Q_n(f; x_0, \ldots, x_n)$ is independent of the order of the points $x_0, \ldots, x_n$.

We show that

$$Q_n(f; x_0, \ldots, x_n) = \frac{\begin{vmatrix} 1 & 1 & \cdots & 1 \\ x_0 & x_1 & \cdots & x_n \\ \cdots\cdots\cdots\cdots\cdots\cdots\cdots \\ \cdots\cdots\cdots\cdots\cdots\cdots\cdots \\ x_0^{n-1} & x_1^{n-1} & \cdots & x_n^{n-1} \\ f(x_0) & f(x_1) & \cdots & f(x_n) \end{vmatrix}}{\begin{vmatrix} 1 & 1 & \cdots & 1 \\ x_0 & x_1 & \cdots & x_n \\ \cdots\cdots\cdots\cdots\cdots\cdots\cdots \\ \cdots\cdots\cdots\cdots\cdots\cdots\cdots \\ x_0^{n-1} & x_1^{n-1} & \cdots & x_n^{n-1} \\ x_0^{n} & x_1^{n} & \cdots & x_n^{n} \end{vmatrix}} = \frac{D\big(f(x)\big)}{D(x^n)}, \text{ say. } (1.1.5)$$

□   If D_r is the co-factor of $f(x_r)$ in $D\big(f(x)\big)$, then

$$D_r = (-1)^{n+r+2} \begin{vmatrix} 1 & 1 & \cdots & 1 & 1 & \cdots & 1 \\ x_0 & x_1 & \cdots & x_{r-1} & x_{r+1} & \cdots & x_n \\ \cdots\cdots\cdots\cdots\cdots\cdots\cdots\cdots\cdots\cdots\cdots\cdots \\ \cdots\cdots\cdots\cdots\cdots\cdots\cdots\cdots\cdots\cdots\cdots\cdots \\ x_0^{n-1} & x_1^{n-1} & \cdots & x_{r-1}^{n-1} & x_{r+1}^{n-1} & \cdots & x_n^{n-1} \end{vmatrix}. \quad (1.1.6)$$

Since the determinants D_r and $D(x^n)$ are van der Monde determinants, we have

$$D(x^n) = \prod_{i<j}(x_j - x_i), \tag{1.1.7}$$

$$D_r = (-1)^{n+r+2}\prod_{i<j}{}'(x_j - x_i), \tag{1.1.8}$$

where in (1.1.8) x_r never occurs. Hence, writing all the factors in these two expressions and cancelling equal factors we get

$$\frac{D(x^n)}{D_r} = (-1)^{n+r+2}(x_n - x_r)(x_{n-1} - x_r)\ldots \tag{1.1.9}$$
$$\times\, (x_{r+1} - x_r)(x_r - x_0)(x_r - x_1)\ldots(x_r - x_{r-1}).$$

Since D_r is the co-factor of $f(x_r)$ in $D\big(f(x)\big)$, we have from the above that

$$\frac{D\big(f(x)\big)}{D(x^n)} = \sum_{r=0}^{n}\frac{f(x_r)D_r}{D(x^n)} = \sum_{r=0}^{n}\frac{f(x_r)}{\prod_{\substack{j=0\\r\neq j}}^{n}(x_r - x_j)} = Q_n(f; x_0, \ldots, x_n),$$
$$\tag{1.1.10}$$

which completes the proof of (1.1.5). $\qquad\square$

From (1.1.5) it follows that

$$Q_n(x^p; x_0, \ldots, x_n) = \begin{cases} 0, & \text{if } p = 0, 1, \ldots, n-1, \\ 1, & \text{if } p = n. \end{cases} \tag{1.1.11}$$

Theorem 1.1.1 *If $f\colon \mathbb{R} \mapsto \mathbb{R}$ is a polynomial of degree at most n, then for every choice of distinct points $x_0, \ldots, x_n$, $Q_n(f; x_0, \ldots, x_n) = 0$ or a_n according, as the degree of f is less than n or equal to n; a_n being the coefficient of x^n in f.*

$\square\qquad$ The proof follows from (1.1.4) and (1.1.11). $\qquad\square$

Theorem 1.1.2 *If $f\colon \mathbb{R} \mapsto \mathbb{R}$, then for every choice of distinct points $x_0, \ldots, x_n$,*

$$Q_n(f; x_0, \ldots, x_n) =$$
$$\frac{y - x_0}{x_n - x_0}Q_n(f; x_0, \ldots, x_{n-1}, y) + \frac{x_n - y}{x_n - x_0}Q_n(f; y, \ldots, x_n), \tag{1.1.12}$$

where y is a point distinct from $x_0, \ldots, x_n$.

$\square\qquad$ From (1.1.1),

$$(x_0 - y)Q_n(f; x_0, \ldots, x_{n-1}, y) =$$
$$Q_{n-1}(f; x_0, \ldots, x_{n-1}) - Q_{n-1}(f; x_1, \ldots, x_{n-1}, y) \tag{1.1.13}$$
$$(y - x_n)Q_n(f; y, x_1, \ldots, x_n) =$$
$$Q_{n-1}(f; y, x_1, \ldots, x_{n-1}) - Q_{n-1}(f; x_1, \ldots, x_n). \tag{1.1.14}$$

Adding (1.1.13) and (1.1.14) and using (1.1.1), we get (1.1.12).   □

Theorem 1.1.3 *Let $f\colon \mathbb{R} \mapsto \mathbb{R}$ and $x_0,\ldots,x_n$ be distinct points. If $y_0,\ldots,y_{r-1}$ is another set of distinct points, each of which is different from $x_0,\ldots,x_n$, rename $x_0,\ldots,x_n,y_0,\ldots,y_{r-1}$ as $z_0,\ldots,z_{n+r}$. Then there are real numbers $\alpha_1,\ldots,\alpha_r$ independent of f such that*

$$Q_n(f;x_0,\ldots,x_n) = \sum_{i=0}^{r}\alpha_i Q_n(f;z_i,\ldots,z_{i+n}) \qquad (1.1.15)$$

and

$$\sum_{i=0}^{r}\alpha_i = 1. \qquad (1.1.16)$$

□   If $r = 1$, the proof follows from Theorem 1.1.2 with $\alpha_0 = (y_0-x_0)/(x_n-x_0)$ and $\alpha_1 = (x_n - y_0)/(x_n - x_0)$.

We suppose that the theorem is true for $r = r_0$ and consider $r_0 + 1$ distinct points $y_0,\ldots,y_{r_0}$ different from $x_0,\ldots,x_n$. Let the points $x_0,\ldots,x_n,y_0,\ldots,y_{r_0-1}$ be renamed as $z_0,\ldots,z_{n+r_0}$. Since the theorem is true for $r = r_0$, there are numbers $\alpha_0,\ldots,\alpha_{r_0}$ such that

$$Q_n(f;x_0,\ldots,x_n) = \sum_{i=0}^{r_0}\alpha_i Q_n(f;z_i,\ldots,z_{i+n}) \qquad (1.1.17)$$

and

$$\sum_{i=0}^{r_0}\alpha_i = 1. \qquad (1.1.18)$$

Now applying (1.1.12) replacing $x_0,\ldots,x_n,y$ by $z_{r_0},\ldots,z_{r_0+n},y_{r_0}$, respectively, we have

$$Q_n(f;z_{r_0},\ldots,z_{r_0+n}) = \frac{y_{r_0} - z_{r_0}}{z_{r_0+n} - z_{r_0}}Q_n(f;z_{r_0},\ldots,z_{r_0+n-1},y_{r_0}) \qquad (1.1.19)$$
$$+ \frac{z_{r_0+n} - y_{r_0}}{z_{r_0+n} - z_{r_0}}Q_n(f;y_{r_0},z_{r_0},\ldots,z_{r_0+n}),$$

and so from (1.1.17) and (1.1.19)

$$Q_n(f;x_0,\ldots,x_n) = \sum_{i=0}^{r_0-1}\alpha_i Q_n(f;z_i,\ldots,z_{i+n}) + \alpha_{r_0}Q_n(f;z_{r_0},\ldots,z_{r_0+n})$$

$$= \sum_{i=0}^{r_0-1}\alpha_i Q_n(f;z_i,\ldots,z_{i+n}) + \alpha_{r_0}\frac{y_{r_0} - z_{r_0}}{z_{r_0+n} - z_{r_0}}Q_n(f;z_{r_0},\ldots,z_{r_0+n-1},y_{r_0})$$

$$+ \alpha_{r_0}\frac{z_{r_0+n} - y_{r_0}}{z_{r_0+n} - z_{r_0}}Q_n(f;y_{r_0},z_{r_0+1},\ldots,z_{r_0+n}). \qquad (1.1.20)$$

Now putting $\beta_i = \alpha_i$, $0 \le i \le r_0-1$, $\beta_{r_0} = \alpha_{r_0} \frac{y_{r_0} - z_{r_0}}{z_{r_0+n} - z_{r_0}}$, $\beta_{r_0+1} = \alpha_{r_0} \frac{z_{r_0+n} - y_{r_0}}{z_{r_0+n} - z_{r_0}}$
we have from (1.1.18)

$$\sum_{i=0}^{r_0+1} \beta_i = \sum_{i=0}^{r_0-1} \alpha_i + \alpha_{r_0} \left[\frac{y_{r_0} - z_{r_0}}{z_{r_0+n} - z_{r_0}} + \frac{z_{r_0+n} - y_{r_0}}{z_{r_0+n} - z_{r_0}} \right] = \sum_{i=0}^{r_0} \alpha_i = 1. \quad (1.1.21)$$

So, renaming the points $z_0, \ldots, z_{r_0+n-1}, y_0, z_{r_0+n}$ as $\omega_0, \ldots, \omega_{n+r_0+1}$ we have from (1.1.20)

$$Q_n(f; x_0. \ldots, x_n) = \sum_{i=0}^{r_0+1} \beta_i Q_n(f; \omega_i, \ldots, \omega_{i+n}). \quad (1.1.22)$$

The relations (1.1.22) and (1.1.21) prove the theorem for $r = r_0 + 1$ and so the theorem is proved by induction. □

Remark. If the points $y_0, \ldots, y_{r-1}$ in Theorem 1.1.3 satisfy

$$\min_{0 \le i \le n} x_i \le y_j \le \max_{0 \le i \le n} x_i, \ j = 0, \ldots, r - 1,$$

then all the α_js are positive.
□ This is true for Theorem 1.1.2 and so the remark follows by the induction above; [145]. □

1.2 General Derivatives of Order n

Let n be a fixed positive integer and let $B = \{a_0, \ldots, a_n\}$ where $a_0, \ldots, a_n$ are distinct real numbers. Let $f \colon \mathbb{R} \mapsto \mathbb{R}$, the *general derivative of f at $x \in \mathbb{R}$ with respect to B* is defined by

$$GD_n f(x, B) = \lim_{h \to 0} n! Q_n(f; x + ha_0, \ldots, x + ha_n), \quad (1.2.1)$$

provided the limit on the right-hand side of (1.2.1) exists. So by (1.1.3)

$$GD_n f(x, B) = \lim_{h \to 0} \frac{n!}{h^n} \sum_{i=0}^{n} \frac{f(x + a_i h)}{\prod_{\substack{j=0 \\ j \neq i}}^{n} (a_i - a_j)}$$

$$= \lim_{h \to 0} \frac{1}{h^n} \sum_{i=0}^{n} A_i f(x + a_i h), \quad (1.2.2)$$

where

$$A_i = \frac{n!}{\prod_{\substack{j=0 \\ j \neq i}}^{n} (a_i - a_j)}. \quad (1.2.3)$$

By (1.1.3) and (1.1.11)

$$\sum_{i=0}^{n} A_i a_i^p = n! \sum_{i=0}^{n} \frac{a_i^p}{\prod_{\substack{j=0 \\ j \neq i}}^{n} (a_i - a_j)} = n! Q_n(x^p; a_0, \dots, a_n) \qquad (1.2.4)$$

$$= \begin{cases} 0, & \text{if } p = 0, \dots, n-1, \\ n!, & \text{if } p = n. \end{cases}$$

Special Cases   **Case I** Let $a_i = i$, $1 = 0, \dots, n$. Then

$$\prod_{\substack{j=0 \\ j \neq i}}^{n} (a_i - a_j) = \prod_{\substack{j=0 \\ j \neq i}}^{n} (i - j)$$

$$= i(i-1) \cdots 1(-1)(-2) \cdots (i - n)$$

$$= (-1)^{(n-i)} i! (n - i)! \qquad (1.2.5)$$

and so

$$A_i = \frac{n!}{(-1)^{(n-i)} i! (n - i)!} = (-1)^{(n-i)} \binom{n}{i}.$$

In this case, we get the *unsymmetric Riemann derivative* of f at x of order n:

$$RD_n f(x) = \lim_{h \to 0} \frac{1}{h^n} \sum_{i=0}^{n} A_i f(x + a_i h) = \lim_{h \to 0} \frac{1}{h^n} \sum_{i=0}^{n} (-1)^{(n-i)} \binom{n}{i} f(x + ih).$$

$$(1.2.6)$$

For $n = 1$, this is

$$RD_1 f(x) = \lim_{h \to 0} \frac{f(x + h) - f(x)}{h}, \qquad (1.2.7)$$

the ordinary derivative of f at x.
For $n = 2$, we get

$$RD_2 f(x) = \lim_{h \to 0} \frac{f(x + 2h) - 2f(x + h) + f(x)}{h^2}. \qquad (1.2.8)$$

Example. If $f(x) = |x|$, then $RD_2 f(x) = 0$ for all x, although $RD_1 f(0)$ does not exist. So, if the nth derivative $RD_n f(x)$ exists, the previous derivatives may not exist.

Case II. Let $a_i = 2i - n$, $i = 0, 1, \dots, n$. Then, as in (1.2.5),

$$\prod_{\substack{j=0 \\ j \neq i}}^{n} (a_i - a_j) = \prod_{\substack{j=0 \\ j \neq i}}^{n} (2i - 2j) = 2^n \prod_{\substack{j=0 \\ j \neq i}}^{n} (i - j)$$

$$= 2^n (-1)^{n-i} i! (n - i)! \qquad (1.2.9)$$

and thus

$$A_i = \frac{n!}{2^n (-1)^{n-i} i! (n - i)!} = \frac{1}{2^n} (-1)^{n-i} \binom{n}{i}. \qquad (1.2.10)$$

In this case, we get the *symmetric Riemann derivative of f at x of order n*:

$$RD_n^s f(x) = \lim_{h \to 0} \frac{1}{h^n} \sum_{i=0}^{n} A_i f(x + a_i h) \tag{1.2.11}$$

$$= \lim_{h \to 0} \frac{1}{(2h)^n} \sum_{i=0}^{n} (-1)^{(n-i)} \binom{n}{i} f(x + 2ih - nh).$$

For $n = 1$, this is

$$RD_1^s f(x) = \lim_{h \to 0} \frac{f(x + h) - f(x - h)}{2h}, \tag{1.2.12}$$

which is the *first symmetric derivative of f at x*.
For $n = 2$, this is

$$RD_2^s f(x) = \lim_{h \to 0} \frac{f(x + 2h) - 2f(x) + f(x - 2h)}{(2h)^2}, \tag{1.2.13}$$

which is the *second symmetric derivative of f at x*.

Example. If $f(x) = |x|$, then $RD_1^s f(x) = 1$ for $x > 0$, $RD_1^s f(x) = -1$ for $x < 0$, $RD_1^s f(0) = 0$ while $RD_2^s f(x) = 0$ for $x \neq 0$, and $RD_2^s f(0) = \infty$.

If the limits in (1.2.1) and consequently in (1.2.6) and (1.2.11) do not exist then the corresponding upper and lower limits are denoted by $\overline{GD}_n f(x; B), \underline{GD}_n f(x; B), \overline{RD}_n f(x), \underline{RD}_n f(x), \overline{RD}_n^s f(x), \underline{RD}_n^s f(x)$, respectively. The unilateral notion of these are obtained by suitably restricting h while taking the limits.

The derivatives $RD_n f(x)$ and $RD_n^s f(x)$ also can be defined with the help of difference operators.

For $RD_n f(x)$, consider the operator, defined by induction

$$\Delta_1(f, x, h) = f(x + h) - f(x) \tag{1.2.14}$$

$$\Delta_k(f, x, h) = \Delta_{k-1}(f, x + h, h) - \Delta_{k-1}(f, x, h), \; k = 2, 3, \cdots.$$

Then,

$$\Delta_n(f, x, h) = \sum_{i=0}^{n} (-1)^{n-i} \binom{n}{i} f(x + ih), \; n = 1, 2, 3, \cdots. \tag{1.2.15}$$

□   For (1.2.15) is true for $n = 1$. Suppose (1.2.15) is true for $n = r$. Then,

$$\Delta_{r+1}(f, x, h) = \Delta_r(f, x+h, h) - \Delta_r(f, x, h)$$

$$= \sum_{i=0}^{r}(-1)^{r-i}\binom{r}{i}f(x+(i+1)h) - \sum_{i=0}^{r}(-1)^{r-i}\binom{r}{i}f(x+ih)$$

$$= \sum_{j=1}^{r+1}(-1)^{r+1-j}\binom{r}{j-1}f(x+jh) - \sum_{i=0}^{r}(-1)^{r-i}\binom{r}{i}f(x+ih)$$

$$= f(x+(r+1)h) + \sum_{j=1}^{r}(-1)^{r+1-j}\left[\binom{r}{j-1} + \binom{r}{j}\right]f(x+jh) + (-1)^{r+1}f(x)$$

$$= f(x+(r+1)h) + \sum_{j=1}^{r}(-1)^{r+1-j}\binom{r+1}{j}f(x+jh) + (-1)^{r+1}f(x)$$

$$= \sum_{j=0}^{r+1}(-1)^{r+1-j}\binom{r+1}{j}f(x+jh),$$

and so (1.2.15) is true for $n = r+1$, proving (1.2.15).   □

So, (1.2.6) can be written as

$$RD_n f(x) = \lim_{h \to 0} \frac{\Delta_n(f, x, h)}{h^n} \tag{1.2.16}$$

where $\Delta_n(f, x, h)$ is given by (1.2.14).

For RD_n^s, consider the operator, defined by induction

$$\Delta_1^s(f, x, h) = f\left(x + \frac{1}{2}h\right) - f\left(x - \frac{1}{2}h\right); \tag{1.2.17}$$

$$\Delta_k^s(f, x, h) = \Delta_{k-1}^s\left(f, x + \frac{1}{2}h, h\right) - \Delta_{k-1}^s\left(f, x - \frac{1}{2}h, h\right), \quad k = 2, 3, \cdots.$$

Then the operators $\Delta_k(\cdot, \cdot, \cdot)$ and $\Delta_k^s(\cdot, \cdot, \cdot)$ have the following relation:

$$\Delta_k^s(f, x, h) = \Delta_k\left(f, x - \frac{1}{2}kh, h\right). \tag{1.2.18}$$

□   In fact, (1.2.18) is true for $k = 1$. Suppose that (1.2.18) is true for $k = r$. Then by (1.2.14), (1.2.17) and by the induction hypothesis

$$\Delta_{r+1}(f, x - \tfrac{1}{2}(r+1)h, h) = \Delta_r(f, x - \tfrac{1}{2}(r-1)h, h) - \Delta_r(f, x - \tfrac{1}{2}(r+1)h, h)$$
$$= \Delta_r^s(f, x + \tfrac{1}{2}h, h) - \Delta_r^s(f, x - \tfrac{1}{2}h, h)$$
$$= \Delta_{r+1}^s(f, x, h),$$

showing that (1.2.18) is true for $k = r+1$ and therefore it is true for all k. □

Hence, by (1.2.18) and (1.2.15)

$$\Delta_n^s(f, x, h) = \sum_{i=0}^{n} (-1)^{n-i} \binom{n}{i} f(x - \tfrac{1}{2}nh + ih). \qquad (1.2.19)$$

Hence, (1.2.11) can be written

$$RD_n^s f(x) = \lim_{h \to 0} \frac{\Delta_n^s(f, x, h)}{h^n}. \qquad (1.2.20)$$

If the limit in (1.2.16) does not exist we write

$$\overline{RD}_n f(x) = \limsup_{h \to 0} \frac{\Delta_n(f, x, h)}{h^n} \text{ and } \underline{RD}_n f(x) = \liminf_{h \to 0} \frac{\Delta_n(f, x, h)}{h^n} \quad (1.2.21)$$

and call them the *upper and lower unsymmetric Riemann derivatives of f at x of order n*, respectively.

If the limit (1.2.20) does not exist, we write

$$\overline{RD}_n^s f(x) = \limsup_{h \to 0} \frac{\Delta_n^s(f, x, h)}{h^n} \text{ and } \underline{RD}_n^s f(x) = \liminf_{h \to 0} \frac{\Delta_n^s(f, x, h)}{h^n} \quad (1.2.22)$$

and call them the *upper and lower symmetric Riemann derivatives of f at x of order n*, respectively.

1.3   Generalized Riemann Derivatives of Order n

Let n be a fixed positive integer and ℓ a nonnegative integer.

Let $S = \{a_0, \ldots, a_{n+\ell}; A_0, \ldots, A_{n+\ell}; L\}$ be a system of real numbers such that $a_i \neq a_j$ for $i \neq j$, $i, j = 0, \ldots, n + \ell$ and $L \neq 0$ satisfying

$$\sum_{i=0}^{n+\ell} A_i a_i^p = \begin{cases} 0, & \text{for } p = 0, \ldots, n - 1, \\ L, & \text{for } p = n. \end{cases} \qquad (1.3.1)$$

If f is defined in some neighbourhood of a point $x \in \mathbb{R}$, then the *generalized Riemann derivative of f at x of order n determined by S* is defined to be

$$GRD_n f(x, S) = \lim_{h \to 0} \frac{n!}{Lh^n} \sum_{i=0}^{n+\ell} A_i f(x + a_i h), \qquad (1.3.2)$$

provided the limit exists.

If in particular $\ell = 0$ and $L = n!$, then the derivative $GRD_n f(x, S)$ becomes the derivative $GD_n f(x, B)$ defined in (1.2.1). To see this (although in this case the expressions (1.2.2) and (1.3.2) are the same), we verify that the constants A_i in (1.3.2) satisfy (1.2.3).

□   If $a_0, \ldots, a_n$ are known, then the relations (1.3.1) determine the A_i uniquely. The coefficient matrix (a_i^p), $0 \le i \le n$, $0 \le p \le n$, of the system of linear equations in (1.3.1) is a van der Monde matrix whose determinant is given by $\det(a_i^p) = \prod_{i<j}(a_j - a_i)$. If c_r^n is the co-factor of a_r^n in $\det(a_i^p)$, then c_r^n is $(-1)^{n+r}$ times a van der Monde determinant and so $c_r^n = (-1)^{n+r} \prod_{i<j}'(a_j - a_i)$ where $\prod'$ means a_r never occurs in $\prod'$. Thus $\dfrac{\det(a_i^p)}{c_r^n} = \prod_{\substack{i=0 \\ i \ne r}}^{n}(a_r - a_i)$.

Hence

$$A_r = L\left[\prod_{\substack{i=0 \\ i \ne r}}^{n}(a_r - a_i)\right]^{-1} = n!\left[\prod_{\substack{i=0 \\ i \ne r}}^{n}(a_r - a_i)\right]^{-1},$$

which is the same as (1.2.3).   □

Thus, in the special case when $\ell = 0$ and $L = n!$ the relations (1.2.4) and (1.3.1) agree and so the derivatives $GRD_n f$ and $GD_n f$, which were defined in (1.3.2) and (1.2.2), respectively, are the same.

The nonnegative number ℓ considered above is called the *excess* if $\ell \ne 0$. The excess terms were first considered by Ash [2]. Excess terms are sometimes necessary in numerical analysis and elsewhere.

If the limit in (1.3.2) does not exist, the *upper and lower generalized Riemann derivatives* of f at x of order n with respect to S are defined by

$$\overline{GRD}_n f(x, S) = \limsup_{h \to 0} \frac{n!}{Lh^n} \sum_{i=0}^{n+\ell} A_i f(x + a_i h)$$

and

$$\underline{GRD}_n f(x, S) = \liminf_{h \to 0} \frac{n!}{Lh^n} \sum_{i=0}^{n+\ell} A_i f(x + a_i h).$$

Now, consider the following system of $n + 1$ linear equations in the $n + 1$ unknowns $A_0, \ldots, A_n$:

$$\sum_{i=0}^{n} A_i = 0,$$

$$\sum_{i=1}^{n} 2^{(i-1)p} A_i = 0 \text{ for } p = 1, \ldots, n - 1, \qquad (1.3.3)$$

$$A_n = 1.$$

The coefficient matrix of the system of linear equations (1.3.3) is nonsingular and so $A_0, \ldots, A_n$ can be determined uniquely.

Note that $\sum_{i=1}^{n} 2^{(i-1)n} A_i \ne 0$. For if $\sum_{i=1}^{n} 2^{(i-1)n} A_i = 0$, then this equation together with the $n - 1$ equations in (1.3.3), $\sum_{i=1}^{n} 2^{(i-1)p} A_i = 0$, $p =$

$1, \ldots, n-1$ form the system of n linear homogeneous equations in the n unknowns $A_1, \ldots, A_n$ and since the determinant of the coefficient matrix of this last system does not vanish, $A_i = 0, i = 1, \ldots, n$, which is a contradiction because $A_n = 1$.

Let

$$L = \sum_{i=1}^{n} 2^{(i-1)n} A_i. \qquad (1.3.4)$$

Now consider the system of real numbers

$$S = \{0, 1, 2, 2^2, \ldots, 2^{n-1}; A_0, \ldots, A_n; L\}. \qquad (1.3.5)$$

Then this system satisfies the conditions (1.3.1) with $\ell = 0$. So, the system (1.3.5) will give a derivative of order n as defined in (1.3.2) with $\ell = 0$. This derivative will be called the *MZ-derivative* after the names of its inventors [106] and will be denoted by $\tilde{D}_n f$. That is, by (1.3.2),

$$\tilde{D}_n f(x) = \lim_{h \to 0} \frac{n!}{h^n \sum_{i=1}^{n} 2^{(i-1)n} A_i} \left[A_0 f(x) + \sum_{i=1}^{n} A_i f(x + 2^{i-1} h) \right]. \qquad (1.3.6)$$

The derivative $\tilde{D}_n f$ also can be defined with the help of difference operators. Consider the difference operator, defined by induction

$$\begin{aligned}
\tilde{\Delta}_1(f, x, h) &= f(x+h) - f(x) && (1.3.7) \\
\tilde{\Delta}_k(f, x, h) &= \tilde{\Delta}_{k-1}(f, x, 2h) - 2^{k-1} \tilde{\Delta}_{k-1}(f, x, h), \quad k = 2, 3, \cdots.
\end{aligned}$$

Then, for any fixed k,

$$\begin{aligned}
\tilde{\Delta}_k(f, x, h) &= \tilde{\Delta}_{k-1}(f, x, 2h) - 2^{k-1} \tilde{\Delta}_{k-1}(f, x, h) \\
&= \tilde{\Delta}_{k-2}(f, x, 2^2 h) - 2^{k-2} \tilde{\Delta}_{k-2}(f, x, 2h) && (1.3.8) \\
&\quad - 2^{k-1} \left[\tilde{\Delta}_{k-2}(f, x, 2h) - 2^{k-2} \tilde{\Delta}_{k-2}(f, x, h) \right] \\
& \cdots \cdots \cdots \cdots \cdots \cdots \cdots \cdots \cdots \cdots \cdots \cdots \cdots \cdots \\
&= \alpha_k f(x + 2^{k-1} h) + \alpha_{k-1} f(x + 2^{k-2} h) + \cdots \\
&\quad + \alpha_1 f(x + h) + \alpha_0 f(x),
\end{aligned}$$

where $\alpha_k = 1$ and α_i depends on i and k only. We show that the α_is satisfy the conditions

$$\sum_{i=0}^{k} \alpha_i = 0, \qquad \sum_{i=1}^{k} 2^{(i-1)s} \alpha_i = 0, \quad \text{for } s = 1, \ldots, k-1. \qquad (1.3.9)$$

☐   We first claim that if f has continuous kth derivative, then

$$\lim_{h \to 0} \frac{\tilde{\Delta}_s(f, x, h)}{h^s} \text{ exists, for } s = 1, \ldots, k. \qquad (1.3.10)$$

Clearly (1.3.10) is true for $s = 1$. So, suppose $1 \leq r < k$ and that (1.3.10) is true for $s = r$. By L'Hôpital's rule

$$\lim_{h \to 0} \frac{(r+1)!}{h^{r+1}} \left[f(x+h) - f(x) - \sum_{j=1}^{r} \frac{h^j}{j!} f^{(j)}(x) \right] = f^{(r+1)}(x),$$

and, so

$$f(x+h) = f(x) + \sum_{j=1}^{r+1} \frac{h^j}{j!} f^{(j)}(x) + o(h^{r+1}). \tag{1.3.11}$$

Since (1.3.8) is true for all k, putting $k = r+1$ in (1.3.8) and substituting in (1.3.11) we have

$$
\begin{aligned}
\tilde{\Delta}_{r+1}(f,x,h) &= \alpha_0 f(x) + \sum_{i=1}^{r+1} \alpha_i f(x + 2^{i-1} h) \tag{1.3.12}\\[2mm]
&= \alpha_0 f(x) + \sum_{i=1}^{r+1} \alpha_i \left[f(x) + \sum_{j=1}^{r+1} \frac{(2^{i-1} h)^j}{j!} f^{(j)}(x) + o(h^{r+1}) \right]\\[2mm]
&= \sum_{i=0}^{r+1} \alpha_i f(x) + \sum_{j=1}^{r+1} \frac{h^j}{j!} f^{(j)}(x) \sum_{i=1}^{r+1} \alpha_i 2^{(i-1)j} + o(h^{r+1}).
\end{aligned}
$$

By the induction hypothesis, (1.3.10) is true for $s = r$ and, therefore, since

$$\frac{\tilde{\Delta}_{r+1}(f,x,h)}{h^{r+1}} = \frac{2^r}{h} \left[\frac{\tilde{\Delta}_r(f,x,2h)}{(2h)^r} - \frac{\tilde{\Delta}_r(f,x,h)}{h^r} \right],$$

we have using (1.3.10) that $\tilde{\Delta}_{r+1}(f,x,h) = o(h^r)$. Hence, from (1.3.12)

$$\sum_{i=0}^{r+1} \alpha_i f(x) + \sum_{j=1}^{r+1} \frac{h^j}{j!} f^{(j)}(x) \sum_{i=1}^{r+1} \alpha_i 2^{(i-1)j} = o(h^r). \tag{1.3.13}$$

Since (1.3.13) is true for all functions f that have a continuous kth derivative, (1.3.13) gives

$$\sum_{i=0}^{r+1} \alpha_i = 0; \qquad \sum_{i=1}^{r+1} \alpha_i 2^{(i-1)j} = 0, \text{ for } j = 1, \ldots, r. \tag{1.3.14}$$

So, (1.3.12) and (1.3.14) imply

$$\tilde{\Delta}_{r+1}(f,x,h) = \frac{h^{r+1}}{(r+1)!} f^{(r+1)}(x) \sum_{i=1}^{r+1} \alpha_i 2^{(i-1)(r+1)} + o(h^{r+1}),$$

showing that

$$\lim_{h \to 0} \frac{\tilde{\Delta}_{r+1}(f,x,h)}{h^{r+1}} = \frac{1}{(r+1)!} f^{(r+1)}(x) \sum_{i=1}^{r+1} \alpha_i 2^{(i-1)(r+1)}. \tag{1.3.15}$$

Clearly (1.3.15) shows that (1.3.10) is also true for $s = r+1$. So, by induction, (1.3.10) is true for all $s = 1, \ldots, k$, whenever f has a continuous kth derivative. Consequently, it follows from (1.3.14) that (1.3.9) is true. □

Now consider $\tilde{\Delta}_n(f, x, h)$ where f is any function. Then from (1.3.8) and (1.3.9)

$$\tilde{\Delta}_n(f, x, h) = \alpha_n f(x + 2^{n-1}h) + \alpha_{n-1} f(x + 2^{n-2}h) + \cdots + \alpha_1 f(x+h) + \alpha_o f(x) \tag{1.3.16}$$

where

$$\sum_{i=0}^{n} \alpha_i = 0; \qquad \sum_{i=1}^{n} \alpha_i 2^{(i-1)s} = 0, \text{ for } s = 1, \ldots, n-1; \qquad \alpha_n = 1. \tag{1.3.17}$$

The system of linear equations (1.3.3) and (1.3.17) are the same. Since the coefficient matrix of this system is nonsingular it will give a unique solution and, therefore, $\alpha_i = A_i$ for $i = 0, 1, \ldots, n$. So, from (1.3.6) and (1.3.16)

$$\tilde{D}_n f(x) = \lim_{h \to 0} \frac{n!}{\sum_{i=1}^{n} 2^{(i-1)n} \alpha_i} \frac{\tilde{\Delta}_n(f, x, h)}{h^n}. \tag{1.3.18}$$

The relation (1.3.18) shows that this derivative $\tilde{D}_n f(x)$ can be defined with the help of the difference operators (1.3.7).

The relation between Δ_k and Δ_k^s being given in (1.2.18), we now show that $\tilde{\Delta}_k$ can be expressed in terms of Δ_k^s.

Theorem 1.3.1 *For each positive integer* $k \geq 2$ *there are constants* $\beta_0, \beta_1, \ldots, \beta_\lambda$ *with* $\beta_\lambda = 1$, *where* $\lambda = 2^{k-1} - k$, *such that*

$$\tilde{\Delta}_k(f, x, h) = \sum_{i=0}^{\lambda} \beta_i \Delta_k^s \left(f, x + \tfrac{1}{2}kh + ih, h\right). \tag{1.3.19}$$

□   Since

$$\Delta_k^s(f, x + \tfrac{1}{2}kh + ih, h) = \sum_{j=0}^{k} (-1)^{k-j} \binom{k}{j} f\left(x + \tfrac{1}{2}kh + ih + jh - \tfrac{1}{2}kh\right)$$

$$= \sum_{j=0}^{k} (-1)^{k-j} \binom{k}{j} f(x + (i+j)h),$$

$\Delta_k^s(f, x + \tfrac{1}{2}kh + ih, h)$ is a linear form in the variables

$$f(x + ih), f(x + (i+1)h), \ldots, f(x + (i+k)h). \tag{1.3.20}$$

Let L_i denote this linear form. Since $k \geq 2$ there are members of (1.3.20) having the form $f(x + 2^{s-1}h)$; s is a positive integer. Let L_i' denote the linear form obtained from L_i by omitting those variables not of the form $f(x + 2^{s-1}h)$.

Considering all the L_i, $i = 0, 1, \ldots, \lambda$, the variables, $2^{k-1}+1$ in number, are by (1.3.20) $f(x+ph)$, $p = 0, 1, \ldots, 2^{k-1}$. Let $L_i'' = L_i - L_i'$. Since the total number of linear forms L_i is $\lambda + 1$, that is also the total number of the linear forms L_i' and L_i''. Considering all the linear forms L_i', $i = 0, 1, \ldots, \lambda$, the variables are

$$f(x + ph), \ p = 0, 1, 2, 2^2, \ldots, 2^{k-1} \tag{1.3.21}$$

of number $k+1$. Therefore, considering all the linear forms L_i'', $i = 0, 1, \ldots, \lambda$, the total number of variables is $(2^{k-1} + 1) - (k + 1) = \lambda$. Since the total number of linear forms L_i'' is $\lambda + 1$, we can find constants $c_0, c_1, \ldots, c_\lambda$, not all of which are zero, such that $c_0 L_0'' + \cdots + c_\lambda L_\lambda'' = 0$ and, hence,

$$c_0 L_0 + c_1 L_1 + \cdots + c_\lambda L_\lambda = c_0 L_0' + c_1 L_0' + \cdots + c_\lambda L_\lambda'. \tag{1.3.22}$$

It follows from (1.3.20) that the variable $f(x + 2^{k-1}h)$ occurs only in L_λ and so only in L_i'. Therefore, from (1.3.22) $c_\lambda \neq 0$, and so again from (1.3.22)

$$\frac{c_0}{c_\lambda} L_0 + \frac{c_1}{c_\lambda} L_1 + \cdots + \frac{c_{\lambda-1}}{c_\lambda} L_{\lambda-1} + L_\lambda = \frac{c_0}{c_\lambda} L_0' + \frac{c_1}{c_\lambda} L_1' + \cdots + \frac{c_{\lambda-1}}{c_\lambda} L_{\lambda-1}' + L_\lambda'. \tag{1.3.23}$$

Now we show that, if f has a kth derivative at x, then for all γ

$$\Delta_k^s(f, x + \gamma h, h) = h^k f^{(k)}(x) + o(h^k). \tag{1.3.24}$$

In fact, we have as in (1.3.11)

$$f(x + h) = f(x) + \sum_{j=1}^{k} \frac{h^j}{j!} f^{(j)}(x) + o(h^k),$$

and so applying the well known relation

$$\sum_{i=0}^{k} (-1)^{k-i} \binom{k}{i} i^p = \begin{cases} 0, & \text{if } p = 0, 1, \ldots, k-1, \\ k!, & \text{if } p = k, \end{cases}$$

we get

$$\begin{aligned}
\Delta_k^s(f, x + \gamma h, h) &= \sum_{i=0}^{k} (-1)^{k-i} \binom{k}{i} f(x + \gamma h + ih - \tfrac{1}{2}kh) \\
&= \sum_{i=0}^{k} (-1)^{k-i} \binom{k}{i} \left[f(x) + \sum_{j=1}^{k} \frac{(\gamma + i - \frac{1}{2}k)^j h^j}{j!} f^{(j)}(x) \right] + o(h^k) \\
&= \sum_{i=0}^{k} (-1)^{k-i} \binom{k}{i} \sum_{j=1}^{k} \frac{(\gamma + i - \frac{1}{2}k)^j h^j}{j!} f^{(j)}(x) + o(h^k)
\end{aligned}$$

$$= \sum_{j=1}^{k} \frac{h^j}{j!} f^{(j)}(x) \sum_{i=0}^{k} (-1)^{k-i} \binom{k}{i} \sum_{\ell=0}^{j} \binom{j}{\ell} i^{\ell} (\gamma - \tfrac{1}{2}k)^{j-\ell} + o(h^k)$$

$$= \sum_{j=1}^{k} \frac{h^j}{j!} f^{(j)}(x) \sum_{\ell=0}^{j} \binom{j}{\ell} (\gamma - \tfrac{1}{2}k)^{j-\ell} \sum_{i=0}^{k} (-1)^{k-i} \binom{k}{i} i^{\ell} + o(h^k)$$

$$= \frac{h^k}{k!} f^{(k)}(x) \sum_{\ell=0}^{k} \binom{k}{\ell} (\gamma - \tfrac{1}{2}k)^{k-\ell} \sum_{i=0}^{k} (-1)^{k-i} \binom{k}{i} i^{\ell} + o(h^k)$$

$$= \frac{h^k}{k!} f^{(k)}(x) k! + o(h^k) = h^k f^{(k)}(x) + o(h^k),$$

proving (1.3.24). So $L_i = h^k f^{(k)}(x) + o(h^k)$ for all i, $i = 0, 1, \ldots, \lambda$. Hence,

$$\frac{c_0}{c_\lambda} L_0 + \frac{c_1}{c_\lambda} L_1 + \cdots + \frac{c_{\lambda-1}}{c_\lambda} L_{\lambda-1} + L_\lambda$$
$$= h^k f^{(k)}(x) \left[\frac{c_0}{c_\lambda} + \frac{c_1}{c_\lambda} + \cdots + \frac{c_{\lambda-1}}{c_\lambda} + 1 \right] + o(h^k). \qquad (1.3.25)$$

Since the variables of $L'_0, L'_1, \ldots, L'_\lambda$ are as in (1.3.21),

$$\frac{c_0}{c_\lambda} L'_0 + \frac{c_1}{c_\lambda} L'_1 + \cdots + \frac{c_{\lambda-1}}{c_\lambda} L'_{\lambda-1} + L'_\lambda$$
$$= b_k f(x + 2^{k-1}h) + b_{k-1} f(x + 2^{k-2}h) + \cdots + b_1 f(x + h) + b_0 f(x) \quad (1.3.26)$$

where the b_i are constants with $b_k = 1$. If f has a kth derivative, then we have as in (1.3.11)

$$f(x + 2^{i-1}h) = f(x) + \sum_{j=1}^{k} \frac{(2^{i-1}h)^j}{j!} f^{(j)}(x) + o(h^k),$$

and so from (1.3.26)

$$\frac{c_0}{c_\lambda} L'_0 + \frac{c_1}{c_\lambda} L'_1 + \cdots + \frac{c_{\lambda-1}}{c_\lambda} L'_{\lambda-1} + L'_\lambda$$
$$= b_0 f(x) + \sum_{i=1}^{k} b_i f(x + 2^{i-1}h)$$
$$= b_0 f(x) + \sum_{i=1}^{k} b_i \left[f(x) + \sum_{j=1}^{k} \frac{(2^{i-1}h)^j}{j!} f^{(j)}(x) + o(h^k) \right]$$
$$= f(x) \sum_{i=0}^{k} b_i + \sum_{j=1}^{k} \frac{h^j}{j!} f^{(j)}(x) \sum_{i=1}^{k} b_i 2^{(i-1)j} + o(h^k). \qquad (1.3.27)$$

From (1.3.23), (1.3.25) and (1.3.27): $\sum_{i=0}^{k} b_i = 0$ and $\sum_{i=1}^{k} b_i 2^{(i-1)j} = 0$ for $j = 1, 2, \ldots, k-1$. Since $b_k = 1$, the b_i satisfy the equations (1.3.17) if α_i in

(1.3.17) is replaced by b_i, $1 = 0, 1, \ldots, k$. So from (1.3.16), the right-hand side of (1.3.26) is $\tilde{\Delta}_k(f, x, h)$. Therefore, from (1.3.26) and (1.3.23)

$$\tilde{\Delta}_k(f, x, h) = \frac{c_0}{c_\lambda}L_0 + \frac{c_1}{c_\lambda}L_1 + \cdots + \frac{c_{\lambda-1}}{c_\lambda}L_{\lambda-1} + L_\lambda. \tag{1.3.28}$$

Putting $\beta_i = c_i/c_\lambda$, $i = 0, 1, \ldots, \lambda - 1$ and $\beta_\lambda = 1$ and noting that $L_i = \Delta_k^s(f, x + \frac{1}{2}kh + ih, h)$ (1.3.28) gives (1.3.19) and completes the proof. □

The above theorem is due to Marcinkiewicz and Zygmund; see Lemma 1.2 of [106, p. 11], the proof of the theorem being simplified here.

1.4   Peano Derivatives

1.4.1   Bilateral Peano Derivatives

Let f be defined in some neighbourhood of the point x on the real line and let k be a positive integer. If there is a polynomial in t, $Q(t) = Q_x(t)$, of degree at most k such that

$$f(x + t) = Q(t) + o(t^k), \text{ as } t \to 0, \tag{1.4.1}$$

then f is said to have a *Peano derivative at x of order k*, and if $a_k/k!$ is the coefficient of t^k in $Q(t)$, then a_k is called the *kth Peano derivative of f at x* and is denoted by $f_{(k)}(x)$.

The Peano derivative, if it exists, is unique.

□   For if $Q_1(t)$ is another polynomial of degree at most k satisfying (1.4.1), then from (1.4.1) $Q(t) = Q_1(t) + o(t^k)$, as $t \to 0$, which implies that for $p = 0, 1, \ldots, k$ we have that $\lim_{t \to 0} \dfrac{Q(t) - Q_1(t)}{t^p} = 0$, and so the polynomials $Q(t)$ and $Q_1(t)$ are identical. □

It is also clear that, if $k \geq 2$ and if $f_{(k)}(x)$ exists, then $f_{(i)}(x)$ exists, $0 \leq i < k$.

□   For if $a_i/i!$ is the coefficient of t^i, then (1.4.1) can be written

$$f(x + t) = S(t) + o(t^i), \text{ as } t \to 0, \text{ where } S(t) = Q(t) - \sum_{j=i+1}^{k} \frac{a_j}{j!}t^j,$$

and S is a polynomial of degree at most i; and so $f_{(i)}(x)$ exists and $a_i = f_{(i)}(x)$. □

If a_0 is the constant term in $Q(t)$, then writing $f_{(0)}(x) = a_0$ we have from (1.4.1)

$$f(x + t) = \sum_{i=0}^{k} \frac{t^i}{i!}f_{(i)}(x) + o(t^k), \text{ as } t \to 0. \tag{1.4.2}$$

So $f_{(0)}(x) = f(x)$ if f is continuous at x, otherwise it is $\lim_{t \to 0} f(x+h)$. Thus if f is continuous at x, then the first Peano derivative $f_{(1)}(x)$ is the first ordinary derivative $f'(x)$.

Supposing the existence of $f_{(k)}(x)$ we write

$$\frac{t^{k+1}}{(k+1)!}\gamma_{k+1}(f;x,t) = f(x+t) - \sum_{i=0}^{k} \frac{t^i}{i!} f_{(i)}(x). \qquad (1.4.3)$$

The *upper and lower Peano derivates of* f at x, of order $k+1$ are obtained by taking the upper and lower limits of $\gamma_{k+1}(f,x,t)$ as $t \to 0$; they are written $\overline{f}_{(k+1)}(x), \underline{f}_{(k+1)}(x)$, respectively.

The *unilateral Peano derivates,* $\overline{f}_{(k+1)}^+(x), \overline{f}_{(k+1)}^-(x), \underline{f}_{(k+1)}^+(x), \underline{f}_{(k+1)}^-(x)$ of order $k+1$ are obtained by suitably restricting t; i.e., $t \to 0+$ or $t \to 0-$. Clearly

$$\overline{f}_{(k+1)}(x) = \max\left[\overline{f}_{(k+1)}^+(x), \overline{f}_{(k+1)}^-(x)\right],$$

$$\underline{f}_{(k+1)}(x) = \min\left[\underline{f}_{(k+1)}^+(x), \underline{f}_{(k+1)}^-(x)\right].$$

If $\overline{f}_{(k+1)}(x) = \underline{f}_{(k+1)}(x)$, then this common value, possibly infinite, is the Peano derivative $f_{(k+1)}(x)$ of f at x of order $k+1$.

Note that if $f_{(k+1)}(x)$ is finite, then this definition of $f_{(k+1)}(x)$ agrees with that given above, as we now show.

□   If $f_{(k+1)}(x)$ is finite, then since $\lim_{t \to 0} \gamma_{k+1}(f;x,t) = f_{(k+1)}(x)$

$$\gamma_{k+1}(f;x,t) = f_{(k+1)}(x) + o(1) \quad \text{as} \quad t \to 0,$$

and so from (1.4.3)

$$f(x+t) = \sum_{i=0}^{k} \frac{t^i}{i!} f_{(i)}(x) + o(t^{k+1}) \quad \text{as} \quad t \to 0,$$

which agrees with (1.4.2). □

The infinite Peano derivative $f_{(k+1)}(x)$, if it exists, is also unique.

□   For if $f_{(k+1)}(x)$ exists, then $f_{(0)}(x), f_{(1)}(x), \ldots, f_{(k)}(x)$ exist and are finite, and from previous observations they are unique and therefore, by definition, $f_{(k+1)}(x)$ is unique. □

Theorem 1.4.1 *If the ordinary kth derivative,* $f^{(k)}(x_0)$, *of* f *at* x_0 *exists, possibly infinite, then the kth Peano derivative,* $f_{(k)}(x_0)$, *also exists and they are equal.*

□   For $k = 1$ the result is trivially true since the first derivative and the first Peano derivative are the same. So, we suppose that $k \geq 2$. Since $f^{(k)}(x_0)$

exists then $f^{(k-1)}(x_0)$ exists, finitely in some neighbourhood of x_0, and then $f^{(k-2)}(x_0)$ exists and is continuous in some neighbourhood of x_0. Let

$$F(t) = f(x_0 + t) - \sum_{i=0}^{k-1} \frac{t^i}{i!} f^{(i)}(x_0),$$

$$G(t) = \frac{t^k}{k!}.$$

Then since $F(0) = F'(0) = \cdots = F^{(k-2)}(0) = 0$ and $G(0) = G'(0) = \cdots = G^{(k-2)}(0) = 0$ for any $t > 0$, there is, by the mean value theorem, an ξ such that $0 < \xi < t$ and

$$\frac{F(t)}{G(t)} = \frac{F^{(k-1)}(\xi)}{G^{(k-1)}(\xi)}. \tag{1.4.4}$$

To see this, for x in some neighbourhood of 0 write

$$\Phi(x) = F(x) + \sum_{i=1}^{k-2} \frac{(t-x)^i}{i!} F^{(i)}(x) - \frac{F(t)}{G(t)} \left[G(x) + \sum_{i=1}^{k-2} \frac{(t-x)^i}{i!} G^{(i)}(x) \right].$$

Observe that $\Phi(0) = \Phi(t) = 0$ and

$$\Phi'(x) = \frac{(t-x)^{k-2}}{(k-2)!} F^{(k-1)}(x) - \frac{F(t)}{G(t)} \frac{(t-x)^{k-2}}{(k-2)!} G^{(k-1)}(x);$$

and so applying the mean value theorem to Φ, $\Phi'(\xi) = 0$, $0 < \xi < t$, which gives (1.4.4).

Since $F^{(k-1)}(\xi) = f^{(k-1)}(x_0 + \xi) - f^{(k-1)}(x_0)$ from (1.4.3) and (1.4.4),

$$\gamma_k(f; x_0, t) = \frac{F(t)}{G(t)} = \frac{f^{(k-1)}(x_0 + \xi) - f^{(k-1)}(x_0)}{\xi}. \tag{1.4.5}$$

Letting $t \to 0+$,

$$\lim_{t \to 0+} \gamma_k(f; x_0, t) = \lim_{\xi \to 0+} \frac{f^{(k-1)}(x_0 + \xi) - f^{(k-1)}(x_0)}{\xi}.$$

Similarly taking $t < 0$ and proceeding as above, we ultimately get

$$\lim_{t \to 0} \gamma_k(f; x_0, t) = \lim_{\xi \to 0} \frac{f^{(k-1)}(x_0 + \xi) - f^{(k-1)}(x_0)}{\xi},$$

completing the proof.   □

Example. The converse of the above theorem is not true except for $k = 1$ when the Peano and ordinary derivative are the same. If n is any positive integer, let

$$f(x) = \begin{cases} x^{n+1} \sin \dfrac{1}{x^n}, & \text{if } x \neq 0, \\ 0, & \text{if } x = 0. \end{cases} \tag{1.4.6}$$

Then f' exists everywhere and

$$f'(x) = \begin{cases} (n+1)x^n \sin \dfrac{1}{x^n} - n \cos \dfrac{1}{x^n}, & \text{if } x \neq 0, \\ 0, & \text{if } x = 0. \end{cases}$$

f' is not differentiable at the origin and so $f^{(k)}(0)$ does not exist for $k \geq 2$. But

$$f(x) = \sum_{i=0}^{n} \frac{x^i}{i!} 0 + x^n \left(x \sin \frac{1}{x^n} \right) \text{ for } x \neq 0.$$

Since $\left(x \sin \dfrac{1}{x^n} \right) \to 0$ as $x \to 0$, it follows from (1.4.1) that $f_{(n)}(0)$ exists and that $f_{(k)}(0) = 0$, for $k = 2, \ldots, n$.

1.4.2 Unilateral Peano Derivatives

Unilateral Peano derivatives can be defined as in section 4.1. We need f to be defined on one side of x, say in some right neighbourhood of x, and (1.4.1) to be replaced by

$$f(x+t) = Q(t) + o(t^k) \text{ as } t \to 0+. \tag{1.4.7}$$

In this case, f is said to have a *right-hand Peano derivative at* x *of order* k, denoted by $Rf_{(k)}(x)$. The relations (1.4.2) and (1.4.3) now take the forms

$$f(x+t) = \sum_{i=0}^{k} \frac{t^i}{i!} Rf_{(i)}(x) + o(t^k), \text{ as } t \to 0+, \tag{1.4.8}$$

$$\frac{t^k}{(k+1!)} \gamma_{k+1}^+(f; x, t) = f(x+t) + \sum_{i=0}^{k} \frac{t^i}{i!} Rf_{(i)}(x), \ t > 0. \tag{1.4.9}$$

The upper and lower limits of $\gamma_{k+1}^+(f, x, t)$ are called the *right upper and lower Peano derivates of* f *at* x *of order* $k + 1$, written $\overline{Rf}_{(k+1)}(x), \underline{Rf}_{(k+1)}(x)$, respectively. If $\overline{Rf}_{(k+1)}(x) = \underline{Rf}_{(k+1)}(x)$, then this common value, possibly infinite, is the right Peano derivative of f at x of order $k + 1$.

Note that the right upper derivate $\overline{f}_{(k+1)}^+(x)$ defined in section 4.1 is, in general, different from $\overline{Rf}_{(k+1)}(x)$ except for $k = 0$, since for $k \neq 0$ we are assuming the existence of $Rf_{(k)}(x)$ only and not $f_{(k)}(x)$ and since the existence of $f_{(k)}(x)$ implies that of $Rf_{(k)}(x)$ and not the converse.

In a similar way one can define left derivatives and derivates that are denoted by $Lf_{(k)}(x)$ and $\overline{Lf}_{(k)}(x), \underline{Lf}_{(k)}(x)$, respectively.

Note that analogues of Theorem 4.1 hold for $Rf_{(k)}(x), Lf_{(k)}(x)$, and that if these two derivatives exist and are equal the bilateral Peano derivative, $f_{(k)}(x)$ may not exist.

Example. Consider for $k \geq 2$,

$$f(x) = \begin{cases} x^{k+1}, & \text{for } x \geq 0, \\ x^{k-1}, & \text{for } x < 0. \end{cases}$$

Then f is continuous and $Rf_{(k)}(0) = 0$, but $f_{(k)}(0)$ does not exist. In fact, if $f_{(k)}(0)$ exists, then $f_{(k-1)}(0)$ also exists. But $\overline{f}^+_{(k-1)}(0) = \underline{f}^+_{(k-1)}(0) = 0$ and $\overline{f}^-_{(k-1)}(0) = \underline{f}^-_{(k-1)}(0) = (k-1)!$.

However, we have the following result:

Theorem 1.4.2 *If $Rf_{(i)}(x) = Lf_{(i)}(x)$ for $i = 1, 2, \ldots, k$, then $f_{(k)}(x)$ exists.*

☐   Clearly the result is true for $k = 1$. Suppose that is true for all k, $1 \leq k \leq r$. Let $Rf_{(i)}(x) = Lf_{(i)}(x)$ for $i = 1, 2, \ldots, r+1$. Since the result holds if $1 \leq k \leq r$, then $f_{(r)}(x)$ exists and $Rf_{(i)}(x) = Lf_{(i)}(x) = f_{(i)}(x)$ for $i = 1, 2, \ldots, r$. Then by (1.4.3) and (1.4.3+)

$$Rf_{(r+1)}(x) = \lim_{t \to 0+} \gamma^+_{r+1}(f, x, t) = \lim_{t \to 0+} \gamma_{r+1}(f, x, t) = \overline{f}^+_{(r+1)}(x) = \underline{f}^+_{(r+1)}(x).$$

Similarly $Lf_{(r+1)}(x) = \overline{f}^-_{(r+1)}(x) = \underline{f}^-_{(r+1)}(x)$. Since $Rf_{(r+1)}(x) = Lf_{(r+1)}(x)$, $f_{(r+1)}(x)$ exists and the result is true for $k = r+1$, completing the proof by induction.   ☐

1.4.3   Peano Boundedness

If in (1.4.1) the condition $o(t^k)$ as $t \to 0$ is replaced by the less restrictive condition $O(t^k)$ as $t \to 0$ then f is said to be *Peano bounded of order k at x*. It is clear that if f is Peano bounded of order k at x, then f need not be Peano differentiable of order k at x. Consider the function f defined in the example in Section 4.1, (1.4.6). We have seen in that example that $f_{(n)}(0)$ exists and $f_{(i)}(0) = 0$ for $i = 1, 2, \ldots, n$. Clearly this function is Peano bounded of order $n+1$ at 0, but the Peano derivative of order $n+1$ at 0 does not exist. However, we have:

Theorem 1.4.3 *If f has finite upper and lower Peano derivates of order k at x, then f is Peano bounded of order k at x.*

☐   Since $\overline{f}_{(k)}(x)$ and $\underline{f}_{(k)}(x)$ are finite, we have that $\gamma_k(f; x, t) = O(1)$ as $t \to 0$. Hence

$$f(x+t) = \sum_{i=1}^{k-1} \frac{t^i}{i!} f_{(i)}(x) + O(t^k) \text{ as } t \to 0.$$

This proves the theorem.   ☐

Theorem 1.4.4 *If f is Peano bounded of order k at x, then f has a finite Peano derivative of order $k - 1$ at x and has finite upper and lower Peano derivates of order k at x.*

□ Since f is Peano bounded of order k at x,

$$f(x + t) = Q(t) + O(t^k) \text{ as } t \to 0, \tag{1.4.6}$$

where $Q(t)$ is a polynomial of degree at most k. Let $Q(t) = \sum_{i=0}^{k} b_i t^i = Q_1(t) + b_k t^k$ say. Then, from (1.4.6)

$$f(x + t) = Q_1(t) + b_k t^k + O(t^k) \text{ as } t \to 0. \tag{1.4.7}$$

Since $O(t^k)$ implies $o(t^{k-1})$ as $t \to 0$, $f_{(k-1)}(x)$ exists and $f_{(i)}(x) = i! b_i$, $i = 0, 1, \ldots, k - 1$. Using (1.4.3) we have from (1.4.7)

$$\gamma_k(f; x, t) = \frac{k!}{t^k} \left[f(x + t) - \sum_{i=0}^{k-1} \frac{t^i}{i!} f_{(i)}(x) \right] = O(1) \text{ as } t \to 0,$$

and, hence, $\overline{f}_{(k)}x$ and $\underline{f}_{(k)}x$ are finite. □

Remark. Unilateral Peano boundedness may similarly be defined and an analogue of Theorem 1.4.3 obtained.

1.4.4 Generalized Peano Derivatives

Let f be continuous in some neighbourhood of $x \in \mathbb{R}$. Then f is said to have a *generalized Peano derivative at x of order n* if there is a nonnegative integer k such that a kth primitive of f in that neighbourhood has a finite Peano derivative at x of order $k+n$. If $f^{(-k)}$ denotes the kth primitive of f and if $(f^{(-k)})_{(n+k)}$ is the $(k+n)$th Peano derivative of $f^{(-k)}$, then $(f^{(-k)})_{(n+k)}(x)$ is called the *generalized Peano derivative of f at x of order n* and is denoted by $f_{[n]}(x)$. That is,

$$f_{[n]}(x) = (f^{(-k)})_{(n+k)}(x). \tag{1.4.8}$$

This definition does not depend on k. For if $k_1 < k_2$ and if (1.4.8) is true for $k = k_1$ then it is also true for $k = k_2$ by L'Hôpital's rule and so k may be taken to be the smallest of all k for which (1.4.8) holds.

Clearly, if $f_{[n]}(x)$, exists then $f_{[i]}(x)$ also exists for $i = 1, 2, \ldots, n - 1$ and $f_{[i]}(x) = (f^{(-k)})_{(k+i)}(x)$.

Now suppose for a fixed positive integer r that $f_{[r-1]}(x)$ exists. Then there is a k such that $f_{[i]}(x) = (f^{(-k)})_{(k+i)}(x)$ for $i = 0, 1, \ldots, r - 1$. Let

$$\Gamma_{k,r}(f;x,h) = \gamma_{k+r}(f^{(-k)};x,h)$$

$$= \frac{(k+r)!}{h^{k+r}}\left[f^{(-k)}(x+h) - \sum_{i=0}^{k+r-1} \frac{h^i}{i!}(f^{(-k)})_{(i)}(x)\right]$$

$$= \frac{(k+r)!}{h^{k+r}}\left[f^{(-k)}(x+h) - \sum_{i=0}^{k-1} \frac{h^i}{i!}f^{(-k+i)}(x) - \sum_{i=0}^{r-1} \frac{h^{k+i}}{(k+i)!}f_{[i]}(x)\right].$$

Define

$$U_k = U_{k,r}(f;x) = \limsup_{h\to 0}\Gamma_{k,r}(f;x,h),$$

$$L_k = L_{k,r}(f;x) = \liminf_{h\to 0}\Gamma_{k,r}(f;x,h).$$

If $k_1 < k_2$ then, by the mean value theorem,

$$\Gamma_{k_2,r}(f;x,h) = \Gamma_{k_1,r}(f;x,\theta h),\ 0 < \theta < 1,$$

and so

$$U_{k_2,r}(f;x) = \limsup_{h\to 0}\Gamma_{k_2,r}(f;x,h) \le \limsup_{h\to 0}\Gamma_{k_1,r}(f;x,h) = U_{k_1,r}(f;x),$$

and similarly $L_{k_2,r}(f;x) \ge L_{k_1,r}(f;x)$. Also for arbitrary k_1, k_2, $L_{k_1,r}(f;x) \le U_{k_2,r}(f;x)$. In fact if $k = \max\{k_1, k_2\}$, then $L_{k_1} \le L_k \le U_k \le U_{k_2}$.
The *upper and lower generalized Peano derivates of f at x of order r* are defined by

$$\overline{f}_{[r]}(x) = \inf_k U_{k,r}(f;x), \quad \underline{f}_{[r]}(x) = \sup_k L_{k,r}(f;x).$$

If $\overline{f}_{[r]}(x) = \underline{f}_{[r]}(x)$, then f is said to have a *generalized Peano derivative*, possibly infinite. It can be shown that if the generalized derivative exists finitely according to this definition, then it exists according to the previous definition; that is, there exists a nonnegative integer k such that (1.4.8) holds.

Now if f is continuous in some neighbourhood of x and if the finite Peano derivative $f_{(r-1)}(x)$ exists, then for any k we have, by the mean value theorem, that $\Gamma_{k,r}(f;x,h) = \gamma_r(f;x,\theta h),\ 0 < \theta < 1$, where γ_r is defined as in (1.4.3). Hence,

$$U_{k,r}(f,x) = \limsup_{h\to 0}\gamma_r(f;x,\theta h)$$

$$\le \limsup_{h\to 0}\gamma_r(f;x,h) = \overline{f}_{(r)}(x),$$

and so $\overline{f}_{[r]}(x) \le \overline{f}_{(r)}(x)$. Similarly, $\underline{f}_{[r]}(x) \ge \underline{f}_{(r)}(x)$ and, hence,

$$\underline{f}_{(r)}(x) \le \underline{f}_{[r]}(x) \le \overline{f}_{[r]}(x) \le \overline{f}_{(r)}(x).$$

Thus it follows that if f is continuous in some neighbourhood of x and if f has a Peano derivative at x of order n, then f has generalized Peano derivative at x of order n. The following example shows that the converse is not true.

Example. Let

$$f(x) = \begin{cases} x^2 \sin \dfrac{1}{x^2}, & \text{if } x \neq 0, \\ 0, & \text{if } x = 0. \end{cases}$$

Then f' exists everywhere, but $f_{(2)}(0)$ does not exist; in fact, $\underline{f}_{(2)}(0) = -2$, $\overline{f}_{(2)}(0) = 2$. We show that $f_{[2]}(0)$ does exist. Let $F(x) = f^{(-1)}(x) = \int_0^x f$. Then

$$F(x) = \int_0^x t^2 \sin(t^{-2}) \, dt = \frac{1}{2} \int_{x^{-2}}^\infty \frac{\sin u}{u^{5/2}} \, du.$$

Let $x^{-2} < X < \infty$. Then, by the second mean value theorem

$$\left| \int_{x^{-2}}^X \frac{\sin u}{u^{5/2}} \, du \right| = \left| x^5 \int_{x^{-2}}^\xi \sin u \, du \right| \leq 2|x^5|$$

and so $|F(x)| \leq |x^5|$; and so, $F_{(i)}(0) = 0$ for $i = 1, 2, 3, 4$, and so, $f_{[i]}(0)$ exists with value zero for $i = 1, 2, 3$.

It may be noted that the existence of $f_{(1)}(x)$ does not imply the continuity of f in a neighbourhood of x and so, in general, $f_{(1)}(x)$ may exist and $f_{[1]}(x)$ may not.

1.4.5   Absolute Peano Derivatives

Let f be defined in some neighbourhood $N(x_0)$ of x_0. The function f is said to have an *absolute Peano derivative* at x_0 if there is a function g and nonnegative integer n such that g has a Peano derivative of order n for all $x \in N(x_0)$ and has a Peano derivative of order $n + 1$ at x_0 and $g_{(n)}(x) = f(x)$, $x \in N(x_0)$. If $g_{(n+1)}(x_0) = A$, then A is called the *absolute Peano derivative* of f at x_0, written $f^*(x_0)$.

Consider the case $n = 0$: If f has an ordinary derivative at x_0, then $f'(x_0) = f_{(1)}(x_0)$, and so take $g = f$ and $f^*(x_0) = f'(x_0)$. However, $f^*(x_0)$ may exist when $f'(x_0)$ does not.

Example. Consider

$$f(x) = \begin{cases} \sin \dfrac{1}{x^2}, & \text{if } x \neq 0, \\ 0, & \text{if } x = 0, \end{cases}$$

and take $g(x) = f^{(-1)}(x_0) = \int_0^x f$. Then by changing the variable and using the second mean value theorem,

$$g(x) = \int_0^x \sin(t^{-2}) \, dt = \frac{1}{2} \int_{x^{-2}}^\infty \frac{\sin u}{u^{3/2}} \, du = \frac{x^3}{2} \int_{x^{-2}}^\xi \sin u \, du,$$

and so $|g(x)| \leq |x^3|$ for all x. This implies that $g_{(1)}(0) = g_{(2)}(0) = 0$. Also $g_{(1)}(x) = \sin \dfrac{1}{x^2}$, $x \neq 0$, and so for all x, $g_{(1)}(x) = f(x)$. Taking $n = 1$ in the above definition, $f^*(0) = g_{(2)}(0) = 0$.

Note that if f has an nth Peano derivative in $N(x_0)$ and has an $(n+1)$st Peano derivative at x_0, then $f_{(n+1)}(x_0)$ is the absolute Peano derivative of $f_{(n)}$ at x_0. From the definition, it follows that if f has an absolute Peano derivative at x_0, then the absolute Peano derivative $f^*(x_0)$ is the Peano derivative of suitable order of some function g and, hence, is a generalized Peano derivative at x_0 of some function. In fact if we use the $\mathcal{C}_r\mathcal{P}$-integral [see Section 8], then since $f(x) = g_{(n)}(x)$ for $x \in N(x_0)$, f is $\mathcal{C}_{n-1}\mathcal{P}$-integrable in $[a, b] \subset N(x_0)$, $x_0 \in (a, b)$. After a finite number of integrations we can get a continuous function F defined on $[a, b]$ such that $F_{(n)}(x) = f(x)$ for $x \in [a, b]$ and $F_{(n+1)}(x_0) = f^*(x_0)$.

1.5 Riemann* Derivatives

1.5.1 Bilateral Riemann* Derivatives

Let x_0 be a fixed point on $\mathbb{R}$ and let $x_1, \ldots, x_k$ be points in some neighbourhood of x_0 such that $0 < |x_0 - x_1| < \cdots < |x_0 - x_k|$. If $Q_k(f; x_0, \ldots, x_k)$ is as defined in (1.1.1) or (1.1.3) and if the iterated limit

$$\lim_{x_k \to x_0} \lim_{x_{k-1} \to x_0} \cdots \lim_{x_1 \to x_0} k! Q_k(f; x_0, \ldots, x_k), \qquad (1.5.1)$$

exists, the last limit being finite or infinite, while the inner limits are all finite, then the limit in (1.5.1) is called the *Riemann* derivative of f at x_0 of order k* and is denoted by $f^*_{(k)}(x_0)$. It is clear that if the last limit in (1.5.1) in finite, then all the inner limits are finite. Clearly $f^*_{(1)}(x_0)$ is the ordinary first derivative $f'(x_0)$. Taking the upper limit, respectively lower limit, at each stage in (1.5.1), we get the definitions of the *upper, respectively lower, Riemann* derivates of f at x_0 of order k*, which are denoted by $\overline{f}^*_{(k)}(x_0)$ and $\underline{f}^*_{(k)}(x_0)$, respectively. In what follows we shall show that this kth order derivative has the properties of the ordinary derivative.

Theorem 1.5.1 *If $k \geq 2$ and if*

$$\lim_{x_{k-1} \to x_0} \cdots \lim_{x_1 \to x_0} Q_k(f; x_0, \ldots, x_k) \qquad (1.5.2)$$

exists finitely for some fixed x_k, then f is continuous at x_0 and

$$\lim_{x_i \to x_0} \cdots \lim_{x_1 \to x_0} Q_i(f; x_0, \ldots, x_i)$$

exists finitely for $i = 1, 2, \ldots, k - 1$.

☐   We have from (1.1.3)

$$Q_k(f; x_0, \ldots, x_k) = \frac{f(x_0)}{(x_0 - x_1) \prod_{j=2}^{k} (x_0 - x_j)} + \frac{f(x_1)}{(x_1 - x_0) \prod_{j=2}^{k} (x_1 - x_j)}$$

$$+ \sum_{i=2}^{k} \frac{f(x_i)}{\prod_{\substack{j=0 \\ j \neq i}}^{k} (x_i - x_j)}$$

$$= \frac{1}{(x_0 - x_1)} \left[\frac{f(x_0)}{\prod_{j=2}^{k} (x_0 - x_j)} - \frac{f(x_1)}{\prod_{j=2}^{k} (x_1 - x_j)} \right]$$

$$+ \sum_{i=2}^{k} \frac{f(x_i)}{\prod_{\substack{j=0 \\ j \neq i}}^{k} (x_i - x_j)}. \tag{1.5.3}$$

The limit in (1.5.2) exists finitely and so all the inner limits are finite and in particular $\lim_{x_1 \to x_0} Q_k(f; x_0, x_1, \ldots, x_k)$ exists finitely. Since the last summation in (1.5.3) tends to a finite limit as $x_1 \to x_0$, we have from (1.5.3) that

$$\lim_{x_1 \to x_0} \frac{1}{(x_0 - x_1)} \left[\frac{f(x_0)}{\prod_{j=2}^{k} (x_0 - x_j)} - \frac{f(x_1)}{\prod_{j=2}^{k} (x_1 - x_j)} \right]$$

exists finitely. Hence,

$$\lim_{x_1 \to x_0} \frac{f(x_1)}{\prod_{j=2}^{k} (x_1 - x_j)} = \frac{f(x_0)}{\prod_{j=2}^{k} (x_0 - x_j)},$$

which shows that $f(x_1) \to f(x_0)$ as $x_1 \to x_0$; that is, f is continuous at x_0.

Now, since f is continuous at x_0, we have from (1.1.3) that

$$\lim_{x_1 \to x_0} Q_{k-1}(f; x_1, x_2, \ldots, x_k) = Q_{k-1}(f; x_0, x_2, \ldots, x_k). \tag{1.5.4}$$

Since, $\big($see (1.1.1)$\big)$,

$$Q_k(f; x_0, x_1, \ldots, x_k) = \frac{Q_{k-1}(f; x_0, x_1, \ldots, x_{k-1}) - Q_{k-1}(f; x_1, x_2, \ldots, x_k)}{x_0 - x_k}, \tag{1.5.5}$$

it follows from (1.5.2) and (1.5.4) that $\lim_{x_1 \to x_0} Q_{k-1}(f; x_0, x_1, \ldots, x_{k-1})$ exists finitely.

If then $k = 2$, the proof is complete; so suppose that $k > 2$.

Calling the last limit $Q_{k-1}(f; x_0, x_0, x_2, \ldots, x_{k-1})$ we conclude that

$$\lim_{x_2 \to x_0} Q_{k-1}(f; x_0, x_2, \ldots, x_k) = Q_{k-1}(f; x_0, x_0, x_3, \ldots, x_k). \tag{1.5.6}$$

From (1.5.4) and (1.5.6),

$$\lim_{x_2 \to x_0} \lim_{x_1 \to x_0} Q_{k-1}(f; x_1, x_2, \ldots, x_k) = Q_{k-1}(f; x_0, x_0, x_3, \ldots .x_k). \tag{1.5.7}$$

From (1.5.2), (1.5.5) and (1.5.7), the $\lim_{x_2 \to x_0} \lim_{x_1 \to x_0} Q_{k-1}(f; x_0, x_1, \ldots, x_{k-1})$ exists finitely. Writing this limit as $Q_{k-1}(f; x_0, x_0, x_0, x_3, \ldots, x_{k-1})$, we conclude that

$$\lim_{x_3 \to x_0} \lim_{x_2 \to x_0} Q_{k-1}(f; x_0, x_2, x_3, \ldots, x_k) = Q_{k-1}(f; x_0, x_0, x_0, x_4, \ldots x_k).$$
(1.5.8)

From the continuity of f at x_0, we have from (1.5.8) that

$$\lim_{x_3 \to x_0} \lim_{x_2 \to x_0} \lim_{x_1 \to x_0} Q_{k-1}(f; x_1, x_2, x_3, \ldots, x_k)$$
$$= Q_{k-1}(f; x_0, x_0, x_0, x_4, \ldots x_k).$$
(1.5.9)

So, by (1.5.2), (1.5.5), and (1.5.9), $\lim_{x_3 \to x_0} \lim_{x_2 \to x_0} \lim_{x_1 \to x_0} Q_{k-1}(f : x_0, x_1, x_2, \ldots, x_{k-1})$ exists finitely.

Now supposing that $\lim_{x_p \to x_0} \lim_{x_{p-1} \to x_0} \cdots \lim_{x_1 \to x_0} Q_{k-1}(f : x_0, x_1, \ldots, x_{k-1})$ exists finitely and writing this limit as $Q_{k-1}(f : (x_0, x_0, \ldots, x_0, x_{p+1}, \ldots, x_{k-1})$ we conclude from this that

$$\lim_{x_{p+1} \to x_0} \cdots \lim_{x_2 \to x_0} Q_{k-1}(f; x_0, x_2, x_3 \ldots, x_{p+1}, x_{p+2}, \ldots, x_k)$$
$$= Q_{k-1}(f; x_0, x_0, x_0 \ldots, x_0, x_{p+2}, \ldots, x_k).$$
(1.5.10)

From the continuity of f at x_0, we conclude from (1.5.10) that

$$\lim_{x_{p+1} \to x_0} \cdots \lim_{x_1 \to x_0} Q_{k-1}(f; x_1, x_2, x_3 \ldots, x_{p+1}, x_{p+2}, \ldots, x_k)$$
$$= Q_{k-1}(f; x_0, x_0, x_0 \ldots, x_0, x_{p+2}, \ldots, x_k).$$
(1.5.11)

From (1.5.2), (1.5.5) and (1.5.11), it follows that

$$\lim_{x_{p+1} \to x_0} \cdots \lim_{x_1 \to x_0} Q_{k-1}(f; x_0, x_1, \ldots, x_{k-1}) \text{ exists finitely.}$$

Continuing this process inductively, we prove that

$$\lim_{x_{k-1} \to x_0} \cdots \lim_{x_1 \to x_0} Q_{k-1}(f; x_0, x_1, \ldots, x_{k-1}) \text{ exists finitely,}$$

thus proving the theorem for $i = k-1$. Then replacing k by $k-1$ and repeating the above argument, we get that

$$\lim_{x_{k-2} \to x_0} \cdots \lim_{x_1 \to x_0} Q_{k-2}(f; x_0, x_1, \ldots, x_{k-2}) \text{ exists finitely.}$$

The proof now follows by a repeated application of this argument. □

Theorem 1.5.2 *Let $k \geq 2$ and suppose that $f^*_{(k)}(x_0)$ exists, finitely or infinitely; then all the previous derivatives $f^*_{(i)}(x_0)$ for $i = 1, 2, \ldots, k-1$ exist finitely.*

□ Since $f^*_{(k)}(x_0)$ exists, it follows that all the inner limits in (1.5.1) are finite and so the condition of Theorem 1.5.1 is satisfied and so, by that theorem, $\lim_{x_i \to x_0} \cdots \ldots \lim_{x_1 \to x_0} i! Q_i(f; x_0, x_1, \ldots, x_i)$ exists finitely for $i = 1, 2, \ldots, k-1$ and this completes the proof. □

1.5.2 Unilateral Riemann* Derivatives

The *unilateral Riemann* derivative and derivates of f at x_0 of order k* can be defined by taking points $x_1, \ldots, x_k$ on the same side of x_0. That is, the *right-hand Riemann* derivative of f at x_0 of order k*, denoted by $Rf^*_{(k)}(x_0)$, is defined to be the iterated limit

$$\lim_{x_k \to x_0+} \cdots \lim_{x_1 \to x_0+} k! Q_k(f; x_0, x_1, \ldots, x_k)$$

where the variables $x_1, \ldots, x_k$ are such that $0 < x_1 - x_0 < x_2 - x_0 < \cdots < x_k - x_0$. Taking lim sup and lim inf instead of lim in the above we get the definitions of the *right-hand upper and lower Riemann* derivates of f at x_0 of order k* that are denoted by $\overline{R}f^*_{(k)}(x_0)$ and $\underline{R}f^*_{(k)}(x_0)$.

The *left-hand Riemann* derivative and derivates of f at x_0 of order k* are defined similarly and they are denoted by $Lf^*_{(k)}(x_0), \overline{L}f^*_{(k)}(x_0)$ and $\underline{L}f^*_{(k)}(x_0)$, respectively.

It can be verified that the analogues Theorem 1.5.1 and Theorem 1.5.2 hold for the one-sided Riemann* derivates; the continuity of f at x_0 in Theorem 1.5.1 is, in this case, only one-sided.

The existence and equality of $Rf^*_{(k)}(x_0)$ and $Lf^*_{(k)}(x_0)$ do not ensure the existence of $f^*_{(k)}(x_0)$. The example given in the case of the Peano derivative can be cited here also. However, the analogue of Theorem 1.4.2 is true.

Theorem 1.5.3 *If $Rf^*_{(i)}(x_0) = Lf^*_{(i)}(x_0)$ for $i = 1, 2, \ldots, k$, then $f^*_{(k)}(x_0)$ exists.*

□ Clearly the theorem is true for $k = 1$. Suppose it is true for $k \leq r$ and let $Rf^*_{(i)}(x_0) = Lf^*_{(i)}(x_0)$ for $i = 1, 2, \ldots, r+1$. Since the theorem is true for $k \leq r$, we conclude that $f^*_{(i)}(x_0)$ exists for $i = 1, 2, \ldots, r$. This shows in particular that f is continuous at x_0. Let $0 < |x_1 - x_0| < |x_2 - x_0| < \cdots < |x_{r+1} - x_0|$. So,

$$\lim_{x_r \to x_0} \cdots \lim_{x_1 \to x_0} Q_r(f; x_0, x_1, \ldots, x_r)$$

exists and, therefore,

$$\lim_{x_{r+1} \to x_0} \cdots \lim_{x_2 \to x_0} Q_r(f; x_0, x_2, \ldots, x_{r+1})$$

exists and so, by the continuity of f at x_0,

$$\lim_{x_{r+1} \to x_0} \cdots \lim_{x_2 \to x_0} \lim_{x_1 \to x_0} Q_r(f; x_1, x_2, \ldots, x_{r+1})$$

exists and, hence,

$$\lim_{x_r \to x_0} \cdots \lim_{x_1 \to x_0} Q_r(f; x_1, x_2, \ldots, x_{r+1}).$$

So, by (1.1.1), $\lim_{x_r \to x_0} \cdots \lim_{x_1 \to x_0} Q_{r+1}(f; x_0, x_1, \ldots, x_{r+1})$ exists. Hence,

$$Rf^*_{(r+1)}(x_0) = \lim_{x_{r+1} \to x_0+} \lim_{x_r \to x_0} \cdots \lim_{x_1 \to x_0} (r+1)! Q_{r+1}(f; x_0, x_1, \ldots, x_{r+1}),$$

$$Lf^*_{(r+1)}(x_0) = \lim_{x_{r+1} \to x_0-} \lim_{x_r \to x_0} \cdots \lim_{x_1 \to x_0} (r+1)! Q_{r+1}(f; x_0, x_1, \ldots, x_{r+1}).$$

Since, by assumption, $Rf^*_{(r+1)}(x_0) = Lf^*_{(r+1)}(x_0)$, we conclude from the last two relations that $f^*_{(r+1)}(x_0)$ exists. The proof is now complete by induction.
$\square$

1.6   Symmetric de la Vallée Poussin Derivatives

1.6.1   Symmetric de la Vallée Poussin Derivative and Symmetric Continuity

Let f be defined in some neighbourhood of x and let k be a fixed positive integer. If there is a polynomial $P(t) = P_x(t)$ of degree at most k such that

$$\tfrac{1}{2}\left[f(x+t) + (-1)^k f(x-t)\right] = P(t) + o(t^k) \text{ as } t \to 0, \qquad (1.6.1)$$

then f is said to have a *kth symmetric de la Vallée Poussin derivative, briefly kth symmetric d.l.V.P. derivative*, or just *kth d.l.V.P. derivative*, at x, and if $a_k/k!$ is the coefficient of t^k in $P(t)$, then a_k is called the *kth d.l.V.P derivative of f at x of order k*, and is denoted by $f^{(s)}_{(k)}(x)$.

It can be shown as in the case of the Peano derivative that if $f^{(s)}_{(k)}(x)$ exists, then it is unique.

It is clear that the polynomial $P(t)$ in (1.6.1) has only even or odd powers of t according as k is even or odd, as we now show.

$\square$ For suppose that $k = 2m$. Then from (1.6.1)

$$f(x+t) + f(x-t) = 2P(t) + o(t^{2m}) \text{ as } t \to 0.$$

Since this is true for positive as well as negative values of t,

$$P(t) = P(-t) + o(t^{2m}) \text{ as } t \to 0$$

and so $P(t)$ has only even powers of t.

If $k = 2m + 1$, then from (1.6.1)

$$f(x+t) - f(x-t) = 2P(t) + o(t^{2m+1}) \text{ as } t \to 0,$$

and so as above

$$P(t) = -P(-t) + o(t^{2m+1}) \text{ as } t \to 0$$

and hence $P(t)$ has only odd powers of t and there is no constant term in $P(t)$ in this case.
$\square$

It is also clear, as we now show, that if $k \geq 2$ and if $f^{(s)}_{(k)}(x)$ exists, then $f^{(s)}_{(k-2)}(x)$ also exists.

☐   For if $a_k/k!$ is the coefficient of t^k in $P(t)$ and if $S(t) = P(t) - \dfrac{a_k}{k!}t^k$, then $S(t)$ is a polynomial of degree at most $k-2$ and from (1.6.1)

$$\tfrac{1}{2}\big[f(x+t) + (-1)^{k-2}f(x-t)\big] = S(t) + o(t^{k-2}) \text{ as } t \to 0,$$

and so $f^{(s)}_{(k-2)}(x)$ exists.   ☐

Therefore, if k is even, then $f^{(s)}_{(0)}(x)$ exists and

$$f^{(s)}_{(0)}(x) = \lim_{h \to 0} \tfrac{1}{2}\big[f(x+t) + f(x-t)\big]. \tag{1.6.2}$$

Definition. A function f is called *symmetrically continuous at x of even order* if

$$f(x+t) + f(x-t) = 2f(x) + o(1) \text{ as } t \to 0, \tag{1.6.3}$$

and *symmetrically continuous at x of odd order* if

$$f(x+t) - f(x-t) = o(1) \text{ as } t \to 0. \tag{1.6.4}$$

In the literature, the condition (1.6.3) is used to mean that f is symmetric at x and condition (1.6.4) is used to mean that f is symmetrically continuous at x; see [168, pp. 24–25]. Clearly, if f is continuous at x, then f is symmetrically continuous at x of both even and odd order. From (1.6.2) it follows that f is symmetrically continuous at x of even order if and only if $f^{(s)}_{(0)}(x) = f(x)$.

To define the upper and lower d.l.V.P. derivates of f at x of order $k+2$, we suppose that $f^{(s)}_{(k)}(x)$ exists and define

$$t\varpi_1(f;x,t) = \tfrac{1}{2}\big[f(x+t) - f(x-t)\big],$$

$$\frac{t^{k+2}}{(k+2)!}\varpi_{k+2}(f;x,t) = \tfrac{1}{2}\big[f(x+t) + (-1)^k f(x-t)\big] - P(t)$$

$$\text{for } k \geq 0, \tag{1.6.5}$$

where

$$P(t) = \begin{cases} \displaystyle\sum_{i=0}^{k/2} \frac{t^{2i}}{(2i)!}f^{(s)}_{(2i)}(x), & \text{if } k \text{ is even,} \\[2.5em] \displaystyle\sum_{i=0}^{(k-1)/2} \frac{t^{2i+1}}{(2i+1)!}f^{(s)}_{(2i+1)}(x), & \text{if } k \text{ is odd.} \end{cases} \tag{1.6.6}$$

The *upper d.l.V.P. derivate of f at x of order $k+2$* is defined by

$$\overline{f}^{(s)}_{(1)}(x) = \limsup_{t \to 0} \varpi_1(f;x,t).$$

$$\overline{f}^{(s)}_{(k+2)}(x) = \limsup_{t \to 0} \varpi_{k+2}(f;x,t), k \geq 0. \tag{1.6.7}$$

Replacing lim sup by lim inf in (1.6.7), we get the definition of the *lower* d.l.V.P. derivate of f at x of order $k+2$, $\underline{f}^{(s)}_{-(k+2)}(x)$. If $\overline{f}^{(s)}_{(k+2)}(x) = \underline{f}^{(s)}_{-(k+2)}(x)$, then this common value is the d.l.V.P. derivative $f^{(s)}_{(k+2)}(x)$, possibly infinite in this case. Note that if $f^{(s)}_{(k+2)}(x)$ is finite, then this definition agrees with the one given above; this we now show.

☐   In fact, if $\overline{f}^{(s)}_{(k+2)}(x) = \underline{f}^{(s)}_{-(k+2)}(x) = \lambda$, finite, then from (1.6.7)

$$\varpi_{k+2}(f;x,t) = \lambda + o(1) \text{ as } t \to 0,$$

and so from (1.6.5)

$$\tfrac{1}{2}\big[f(x+t) + (-1)^k f(x-t)\big] = \frac{t^{k+2}}{(k+2)!}\varpi_{k+2}(f;x,t) + P(t)$$

$$= P(t) + \frac{t^{k+2}}{(k+2)!}\lambda + o(t^{k+2}) \text{ as } t \to 0.$$

Since from (1.6.6), $P(t)$ is a polynomial of degree at most k, $P(t) + \frac{t^{k+2}}{(k+2)!}\lambda$ is a polynomial of degree at most $k+2$ with $\lambda/(k+2)!$ as the coefficient of t^{k+2}, and so the result follows.   ☐

As in the case of the Peano derivative, it can be shown that the d.l.V.P. derivative, if it exists, is unique.

It may be noted that if $f^{(s)}_{(k)}(x)$ exists then $f^{(s)}_{(k-1)}(x)$ may not exist.

Example.

$$f(x) = \begin{cases} x\cos\dfrac{1}{x}, & \text{if } x \neq 0, \\ 0, & \text{if } x = 0. \end{cases}$$

Since $f(t)+f(-t) = 0$, $f^{(s)}_{(k)}(0)$ exists for all even k. But, $f(t)-f(-t) = 2t\cos\dfrac{1}{t}$ and so $f^{(s)}_{(1)}(0)$ does not exist and, therefore, $f^{(s)}_{(k)}(0)$ cannot exists if k is odd.

The d.l.V.P. derivatives of order 1 and 2 are respectively called the *first and second symmetric derivatives* and they coincide with the symmetric Riemann derivatives of order 1 and 2, respectively (cf. (1.2.12) and (1.2.13)), when f is continuous at x.

1.6.2   Smoothness

It has been noted that if $f^{(s)}_{(k)}(x)$ exists, then $f^{(s)}_{(k-1)}(x)$ might not exist, although $f^{(s)}_{(k-2)}(x)$ must exist finitely. A property fitted between $f^{(s)}_{(k-2)}(x)$ and $f^{(s)}_{(k)}(x)$ is the property of smoothness of f of order k at x.

Let f be defined in some neighbourhood of x and let $f^{(s)}_{(k-2)}(x)$ exist finitely. Define

$$\overline{S}_k(f;x) = \limsup_{h \to 0+} h\varpi_k(f;x,h),$$

$$\underline{S}_k(f;x) = \liminf_{h \to 0+} h\varpi_k(f;x,h).$$

Then $\overline{S}_k(f;x)$ and $\underline{S}_k(f;x)$ are respectively called the *upper and lower index of d.l.V.P. smoothness* of f of order k at x. If $\overline{S}_k(f;x) = \underline{S}_k(f;x)$, the common value, denoted by $S_k(f;x)$, is called the *index of d.l.V.P. smoothness* of f of order k at x. If $S_k(f;x)$ exists and $S_k(f;x) = 0$, then f is said to be *d.l.V.P. smooth* or simply *smooth*, of order k at x. If f is smooth of order 2 at x, then f is said to be *smooth* at x, omitting the order [?, 169]. If, however,

$$-\infty < \underline{S}_k(f;x) \le \overline{S}_k(f;x) < \infty,$$

then f is said to be *quasi-smooth of order k at x*; for the case $k = 2$, called just *quasi-smoothness*, see [168, p. 124]. In other words, f is said to be smooth or quasi-smooth of order k at x accordingly as $h\varpi_k(f;x,h) = o(1)$ or $h\varpi_k(f;x,h) = O(1)$ as $h \to 0$.

Theorem 1.6.1 *If $\overline{f}^{(s)}_{(k)}(x)$ and $\underline{f}^{(s)}_{(k)}(x)$ are finite, then f is smooth of order k at x.*

□   Since $\overline{f}^{(s)}_{(k)}(x)$ and $\underline{f}^{(s)}_{(k)}(x)$ are finite, $\varpi_k(f;x,t)$ remains bounded as $t \to 0$ and so $\lim_{t \to 0} t\varpi_k(f;x,t) = 0$, which shows that $\overline{S}_k(f;x) = \underline{S}_k(f;x) = 0$ and, hence, f is smooth of order k at x.   □

Corollary 1.6.2 *If f is smooth of order k at x, then f is smooth of order $k - 2$ at x.*

□   If f is smooth of order k at x, then by definition $f^{(s)}_{(k-2)}(x)$ exists finitely and so by the above, f is smooth of order $k - 2$ at x.   □

Remark. If f is smooth of order k at x, then f may not be smooth of order $k - 1$ at x because $f^{(s)}_{(k-3)}(x)$ may not exist.

1.6.3   de la Vallée Poussin Boundedness

If in (1.6.1) the condition $o(t^k)$ as $t \to 0$ is replaced by the less restrictive condition $O(t^k)$ as $t \to 0$, then f is said to be *d.l.V.P. bounded* at x of order k.

Example. Consider the function

$$f(x) = \begin{cases} x^{2k} \sin \dfrac{1}{x^{2k}}, & \text{if } x \ne 0, \\ 0, & \text{if } x = 0. \end{cases}$$

Then $\frac{1}{2}\left[f(t)+(-1)^{2k}f(-t)\right]=t^{2k}\sin(1/x^{2k})=O(t^{2k})$ as $t\to 0$. Since $O(t^{2k})$ as $t\to 0$ implies $o(t^{2k-2})$ as $t\to 0$, $f^{(s)}_{(2i)}(0)$ exists and $f^{(s)}_{(2i)}(0)=0$ for $i=0,1,\ldots,k-1$. Also $f^{(s)}_{(2k)}(0)$ does not exist, but f is d.l.V.P. bounded of order $2k$ at 0.

However, we have:

Theorem 1.6.3 *If f is d.l.V.P. bounded at x of order k, then f has a finite d.l.V.P. derivative at x of order $k-2$ and has finite upper and lower d.l.V.P. derivates of order k at x.*

◻   Let k be even, say $k=2m$. Since f is d.l.V.P. bounded at x of order $2m$,

$$\frac{1}{2}\left[f(x+t)+f(x-t)\right]=P(t)+O(t^{2m})\text{ as }t\to 0,\qquad(1.6.8)$$

where $P(t)$ is a polynomial of degree at most $2m$. It can be shown that, in this case, $P(t)$ has only even powers of t. So, let $P(t)=\sum_{i=0}^{m}b_{2i}t^{2i}$. Then writing $P_1(t)=\sum_{i=0}^{m-1}b_{2i}t^{2i}$, we have from (1.6.8) that:

$$\frac{1}{2}\left[f(x+t)+f(x-t)\right]=P_1(t)+b_{2m}t^{2m}+O(t^{2m})=P_1(t)+O(t^{2m})\text{ as }t\to 0.$$
$$(1.6.9)$$

Since $O(t^{2m})$ as $t\to 0$ implies $o(t^{2m-2})$ as $t\to 0$, we have from (1.6.9) that $f^{(s)}_{(2m-2)}(x)$ exists and $f^{(s)}_{(2i)}(x)=(2i)!b_{2i}$ for $i=0,1,\ldots,m-1$. Using (1.6.5), we have from (1.6.9)

$$\varpi_{2m}(f;x.t)=\frac{(2m)!}{t^{2m}}\left[\frac{f(x+t)+f(x-t)}{2}-\sum_{i=0}^{m-1}\frac{t^{2i}}{(2i)!}f^{(s)}_{(2i)}(x)\right]=O(1)\text{ as }t\to 0,$$

and, hence, $\overline{f}^{(s)}_{(2m)}(x)$ and $\underline{f}^{(s)}_{(2m)}(x)$ are finite. This proves the theorem for even k. The proof for odd k is similar.   ◻

Theorem 1.6.4 *If f has finite upper and lower d.l.V.P. derivates at x of order k, then f is d.l.V.P. bounded at x of order k.*

◻   Since $\overline{f}^{(s)}_{(k)}(x)$ and $\underline{f}^{(s)}_{(k)}(x)$ are finite, we have $\varpi_k(f;x.t)=O(1)$ as $t\to 0$ and hence from (1.6.5)

$$\frac{1}{2}\left[f(x+t)+(-1)^k f(x-t)\right]=P(t)+O(t^k)\text{ as }t\to 0,$$

where $P(t)$ is the polynomial defined in (1.6.6). This proves that f is d.l.V.P. bounded at x of order k.   ◻

Theorem 1.6.5 *If f is d.l.V.P. bounded at x of order k, then f is smooth at x of order k.*

◻   The proof follows from Theorem 1.6.3 and Theorem 1.6.4.   ◻

1.7 Symmetric Riemann* Derivatives

Let x be a fixed point on $\mathbb{R}$ and let f be defined in some neighbourhood of x. Let n be a fixed positive even integer, $n = 2k$ say, and consider $0 < h_1 < h_2 < \cdots < h_k$. If the iterated limit

$$\lim_{h_k \to 0} \cdots \lim_{h_1 \to 0} n! Q_n(f; x-h_k, x-h_{k-1}, \cdots, x-h_1, x, x+h_1, \cdots, x+h_k) \quad (1.7.1)$$

exists, the last limit being finite or infinite, while the inner limits are finite, and where $Q_n(f; x-h_k, x-h_{k-1}, \cdots, x-h_1, x, x+h_1, \cdots, x+h_k)$ is the divided difference of f at the points $x-h_k, x-h_{k-1}, \cdots, x-h_1, x, x+h_1, \cdots, x+h_k$ as defined in (1.1.1), then this limit is called the *symmetric Riemann** derivative of f at x of order $n, n = 2k$.

If n is a positive odd integer, $n = 2k - 1$ say, consider $0 < h_1 < h_2 < \cdots < h_k$. If the iterated limit

$$\lim_{h_k \to 0} \cdots \lim_{h_1 \to 0} n! Q_n(f; x-h_k, x-h_{k-1}, \cdots, x-h_1, x+h_1, \cdots, x+h_k) \quad (1.7.2)$$

exists, the last limit being finite or infinite, while the inner limits are finite, then this limit is called the *symmetric Riemann* derivative of f at x of order* $n, n = 2k - 1$. The symmetric Riemann* derivative of f at x of order n is written $f_{(n)}^{*(s)}(x)$.

Clearly $f_{(1)}^{*(s)}(x)$ and $f_{(2)}^{*(s)}(x)$ are respectively the first and second symmetric derivatives of f at x considered in (1.2.12) and (1.2.13). Taking "lim sup" and "lim inf" instead of "lim" at each stage of (1.7.1) or (1.7.2) we get the corresponding *upper* and *lower symmetric Riemann* derivates of f at x of order n, denoted by $\overline{f}_{(n)}^{*(s)}(x)$ and $\underline{f}_{(n)}^{*(s)}(x)$, respectively.

1.8 Cesàro Derivatives

The Cesàro derivative of order r of a function is defined with the help of an integral known as the Cesàro–Perron integral of order $r - 1$. Interestingly, the Cesàro–Perron integral of order $r - 1$ is defined with the help of the Cesàro derivative of order $r - 1$. Thus, the definition of order r is obtained step-by-step starting from order 0, the Cesàro–Perron integral of order 0 being the special Denjoy integral or what is also the Perron integral.

1.8.1   Cesàro Continuity and Cesàro Derivative

The Cesàro continuity and Cesàro derivative of order 0, briefly called C_0-continuity and C_0-derivative, are respectively ordinary continuity and the ordinary first derivative. The Cesàro–Perron integral of order 0, briefly the $C_0 P$-integral, is the special Denjoy integral, or equivalently the Perron integral or more recently the Henstock integral; see [10, 28, 29]. Starting from this we get the definition of C_r-continuity and C_r-derivative as follows.

Suppose that the definitions of C_{r-1}-continuity, the C_{r-1}- derivative and the $C_{r-1} P$-integral are known. Also suppose that if f is $C_{r-1} P$-integrable, then Qf is also $C_{r-1} P$-integrable for every polynomial Q and that the integration by parts formula holds; this is known when $r = 1$, the C_0 case.

Let f be $C_{r-1} P$-integrable on $[a, b]$, then for $x, x + h$ in $[a, b]$ write

$$C_0(f; x, x + h) = f(x + h),$$

$$C_r(f; x, x + h) = \frac{r}{h^r}(C_{r-1}P)\!\int_x^{x+h} (x + h - t)^{r-1} f(t)\, dt, \quad r \geq 1, \quad (1.8.1)$$

where $(C_{r-1}P)$ indicates that the integral is a $C_{r-1} P$-integral. The quantity $C_r(f; x, x + h)$ is called the *r-th Cesàro mean of f on the interval with endpoints* x *and* $x + h$.

The *right upper and lower C_r-limits of f at x* are defined by

$$\overline{C}_r^+ f(x) = \limsup_{h \to 0+} C_r(f; x, x + h), \qquad \underline{C}_r^+ f(x) = \liminf_{h \to 0+} C_r(f; x, x + h),$$

respectively.

The *left C_r-limits*, $\overline{C}_r^- f(x)$ and $\underline{C}_r^- f(x)$, are defined by considering $h \to 0-$. Define the *upper and lower C_r-limits of f at x* by

$$\overline{C}_r f(x) = \max\{\overline{C}_r^+ f(x), \overline{C}_r^- f(x)\}, \quad \underline{C}_r f(x) = \min\{\underline{C}_r^+ f(x), \underline{C}_r^- f(x)\}.$$

If $\overline{C}_r f(x) = \underline{C}_r f(x)$, the common value, denoted by $C_r f(x)$, is called the *C_r-limit of f at x*. If $C_r f(x)$ exists and is equal to $f(x)$, then f is said to be *C_r-continuous at x*.

The *right upper and lower C_r-derivates of f at x* are defined by

$$\overline{C_r D}^+ f(x) = \limsup_{h \to 0+} \frac{r+1}{h}\left[C_r(f; x, x + h) - f(x)\right],$$

$$\underline{C_r D}^+ f(x) = \liminf_{h \to 0+} \frac{r+1}{h}\left[C_r(f; x, x + h) - f(x)\right],$$

respectively. The left C_r-derivates of f at x, $\overline{C_r D}^- f(x)$ and $\underline{C_r D}^- f(x)$, are defined by considering $h \to 0-$. The *upper and lower C_r-derivates of f at x* are defined, respectively, by

$$\overline{C_r D} f(x) = \max\{\overline{C_r D}^+ f(x), \overline{C_r D}^- f(x)\},$$

$$\underline{C_r D} f(x) = \min\{\underline{C_r D}^+ f(x), \underline{C_r D}^- f(x)\}.$$

A function M is called a C_r-*major function of* f *on* $[a, b]$ if

(i) M is C_r-continuous,

(ii) $M(a) = 0$,

(iii) $\underline{C_r D} M(x) \geq f(x)$ for all x in $[a, b]$,

(iv) $\underline{C_r D} M(x) > -\infty$ for all x in $[a, b]$,

where, at the end points a and b, one-sided continuity and one-sided derivates will be considered. A function m is a C_r-*minor function of* f *on* $[a, b]$ if $-m$ is a C_r-major function of $-f$ on $[a, b]$.

If $\{M\}$ and $\{m\}$ denote the family of C_r-major and C_r-minor functions respectively then it can be shown that $M - m$ is increasing on $[a, b]$ for any $M \in \{M\}$ and $m \in \{m\}$. Since $M(a) - m(a) = 0$, it follows that $M(b) - m(b) \geq 0$ and so $\sup_{m \in \{m\}} m(b) \leq \inf_{M \in \{M\}} M(b)$. If $\sup_{m \in \{m\}} m(b) = \inf_{M \in \{M\}} M(b) \neq \pm\infty$, then f is said to be $C_r P$-*integrable* and this common value is the $C_r P$- integral of f on $[a, b]$ and is denoted by $(C_r P) \int_a^b f(t) \, \mathrm{d}t$.

It is known that if f is $C_{r-1} P$-integrable, then f is $C_r P$-integrable and the integrals are equal. Further, the $C_r P$-integral admits the following integration by parts formula:

Theorem 1.8.1 *Let* f *be* $C_r P$-*integrable on* $[a, b]$ *and let* $F(x) = (C_r P) \int_a^x f(t) \, \mathrm{d}t$, $x \in [a, b]$. *If* G *is an* r *times repeated integral of a function of bounded variation in* $[a, b]$, *then* fG *is* $C_r P$-*integrable on* $[a, b]$ *and*

$$(C_r P) \int_a^b fG = FG \Big|_a^b - (C_{r-1} P) \int_a^b FG'.$$

Lemma 1.8.2 *If* $r \geq 1$ *and* f *is* $C_{r-1} P$-*integrable in some neighbourhood of* 0, *then for small* $|h| \neq 0$

$$\int_0^h (h - t)^r f(t) \, \mathrm{d}t = r \int_0^h \int_0^\xi (\xi - t)^{r-1} f(t) \, \mathrm{d}t \, \mathrm{d}\xi. \tag{1.8.2}$$

$\square$ We have, integrating by parts successively,

$$\int_0^h (h - t)^r f(t) \, \mathrm{d}t = r! \int_0^h \int_0^{\xi_1} \cdots \int_0^{\xi_r} f(\xi) \, \mathrm{d}\xi \, \mathrm{d}\xi_r \cdots \mathrm{d}\xi_1 \tag{1.8.3}$$

and similarly

$$\int_0^{\xi_1} (\xi_1 - t)^{r-1} f(t) \, \mathrm{d}t = (r-1)! \int_0^{\xi_1} \int_0^{\xi_2} \cdots \int_0^{\xi_r} f(\xi) \, \mathrm{d}\xi \, \mathrm{d}\xi_r \cdots \mathrm{d}\xi_2. \tag{1.8.4}$$

Integrating (1.8.4), we have

$$r \int_0^h \int_0^{\xi_1} (\xi_1 - t)^{r-1} f(t) \, \mathrm{d}t \, \mathrm{d}\xi_1 = r! \int_0^h \int_0^{\xi_1} \cdots \int_0^{\xi_r} f(\xi) \, \mathrm{d}\xi \, \mathrm{d}\xi_r \cdots \mathrm{d}\xi_1, \tag{1.8.5}$$

and so (1.8.2) follows from (1.8.3) and (1.8.5). $\qquad \square$

Theorem 1.8.3 *If $r \geq 1$ and f is $C_{r-1}P$-integrable in $[a, b]$, then for $a \leq x < b$*

(i) $\underline{C}_r^+ f(x) \leq \underline{C}_{r+1}^+ f(x) \leq \overline{C}_{r+1}^+ f(x) \leq \overline{C}_r^+ f(x)$;

(ii) $\underline{C}_r D^+ f(x) \leq \underline{C}_{r+1} D^+ f(x) \leq \overline{C}_{r+1} D^+ f(x) \leq \overline{C}_r D^+ f(x)$,

with similar results for the left limits and left derivates.

☐   (i) We may suppose that $\overline{C}_r^+ f(x) < \infty$ and that $x = 0$.

Choose any λ such that $\overline{C}_r^+ f(x) < \lambda < \infty$. Then there is a $\delta > 0$ such that $C_r(f; 0.h) < \lambda$ for $0 < h < \delta$. So,

$$r \int_0^\xi (\xi - t)^{r-1} f(t)\, dt < \lambda \xi^r \quad \text{for } 0 < \xi < \delta.$$

Hence,

$$r \int_0^h \int_0^\xi (\xi - t)^{r-1} f(t)\, dt\, d\xi \leq \lambda \frac{h^{r+1}}{r+1} \quad \text{for } 0 < h < \delta.$$

So, by, Lemma 1.8.2

$$\int_0^h (h - t)^r f(t)\, dt \leq \lambda \frac{h^{r+1}}{r+1} \quad \text{for } 0 < h < \delta,$$

which gives

$$C_{r+1}(f; 0.h) = \frac{r+1}{h^{r+1}} \int_0^h (h - t)^r f(t)\, dt \leq \lambda \quad \text{for } 0 < h < \delta.$$

Hence, $\overline{C}_{r+1}^+ f(x) \leq \lambda$. Since λ is arbitrary, the last inequality follows. The first inequality is proved in a similar manner.

(ii) As in (i), we suppose that $\overline{C}_r D^+ f(0) < \lambda < \infty$. We also suppose that $f(0) = 0$. Then there is a $\delta > 0$ such that $\frac{r+1}{t} C_r(f; 0, t) < \lambda$ for $0 < t < \delta$. Hence,

$$\int_0^t (t - \xi)^{r-1} f(\xi)\, d\xi < \lambda \frac{t^{r+1}}{r(r+1)} \quad \text{for } 0 < t < \delta.$$

Hence,

$$r \int_0^h \int_0^t (t - \xi)^{r-1} f(\xi)\, d\xi\, dt \leq \lambda \frac{h^{r+2}}{(r+1)(r+2)} \quad \text{for } 0 < h < \delta.$$

So, by Lemma 1.8.2

$$\int_0^h (h - \xi)^r f(\xi)\, d\xi \leq \lambda \frac{h^{r+2}}{(r+1)(r+2)} \quad \text{for } 0 < h < \delta,$$

which gives

$$\frac{r+2}{h}C_{r+1}(f;0,h) \leq \lambda \text{ for } 0 < h < \delta,$$

and so $\overline{C_{r+1}D}^+ f(0) \leq \lambda$, which completes the proof of (ii) as in (i). $\square$

If $r = 0$, we have the following theorem. Remember that for $r = 0$ the notation C_0 refers to the ordinary limits and derivates, and the special Denjoy integral.

Theorem 1.8.4 *If f is C_0P-integrable in $[a, b]$, then for $a \leq x \leq b$*

(i) $\underline{C_0^+} f(x) \leq \underline{C_1^+} f(x) \leq \overline{C_1}^+ f(x) \leq \overline{C_0}^+ f(x);$

(ii) $\underline{C_0D}^+ f(x) \leq \underline{C_1D}^+ f(x) \leq \overline{C_1D}^+ f(x) \leq \overline{C_0D}^+ f(x).$

$\square$ (i) Suppose $x = 0$ and $\overline{C_0}^+ f(0) < \infty$. Choose $\overline{C_0}^+ f(0) < \lambda < \infty$. Then there exists $\delta > 0$ such that $C_0(f; 0, t) < \lambda$ for $0 < t < \delta$. Since $C_0(f; 0, t) = f(t)$, $f(t) < \lambda$ for $0 < t < \delta$ and so

$$C_1(f; 0, h) = \frac{1}{h} \int_0^h f \leq \lambda \text{ for } 0 < h < \delta.$$

Letting $h \to 0$, $\overline{C_1}^+ f(0) \leq \lambda$ and the result follows as above.

(ii) Suppose that $x = 0 = f(x)$ and $\overline{C_0D}^+ f(0) < \infty$. Choose $\overline{C_0D}^+ f(0) < \lambda < \infty$. Then there is a $\delta > 0$ such that $\frac{1}{t}C_0(f; 0, t) < \lambda$ for $0 < t < \delta$. Hence, $f(t) < \lambda t$ for $0 < t < \delta$ and, hence,

$$\int_0^t f \leq \lambda \frac{t^2}{2}, \text{ for } 0 < t < \delta,$$

which gives $\frac{2}{t}C_1(f; 0, t) \leq \lambda$ for $0 < t < \delta$. Hence, $\overline{C_1D}^+ f(0) \leq \lambda$. The rest is clear. $\square$

Corollary 1.8.5 *Let f be $C_{r-1}P$-integrable, $r \geq 1$. If f is C_r-continuous at x, then f is C_{r+1}-continuous at x and, if $C_rDf(x)$ exists, then $C_{r+1}Df(x)$ exists and equals $C_rDf(x)$.*

Corollary 1.8.6 *Let f be C_0P-integrable. If f is continuous at x, then f is C_1-continuous at x and, if the ordinary derivative $f'(x)$ exists, then $C_1Df(x)$ exists and equals $f'(x)$.*

Theorem 1.8.7 *Let f be $C_{r-1}P$-integrable, $r \geq 1$ and let Φ be its indefinite $C_{r-1}P$-integral. Then the upper and lower C_{r-1}-derivates of Φ at x are, respectively, the upper and lower C_r-limits of f at x.*

☐   Integrating by parts, we have

$$\frac{r}{h^r}\int_x^{x+h}(x+h-t)^{r-1}f(t)\,dt = \frac{r}{h}\left[\frac{1}{h^{r-1}}(x+h-t)^{r-1}\int_x^t f\Big|_{t=x}^{x+h}\right]$$

$$+\frac{r-1}{h^{r-1}}\int_x^{x+h}(x+h-t)^{r-2}\int_x^t f(\xi)\,d\xi\,dt$$

$$= \frac{r}{h}\left[\frac{r-1}{h^{r-1}}\int_x^{x+h}(x+h-t)^{r-2}\big(\Phi(t)-\Phi(x)\big)\,dt\right]$$

$$= \frac{r}{h}\left[C_{r-1}(\Phi;x,x+h)-\Phi(x)\right],$$

and so from (1.8.1)

$$C_r(f;x,x+h) = \frac{r}{h}\left[C_{r-1}(\Phi;x,x+h)-\Phi(x)\right],$$

which proves the theorem.   ☐

Corollary 1.8.8 *A function f is C_r-continuous at x if and only if $f(x)$ is the C_{r-1}-derivate of its indefinite $C_{r-1}P$-integral.*

1.8.2   Cesàro Boundedness

A function f is said to be *Cesàro bounded of order* r, $r \geq 1$, briefly C_r-*bounded at x* if

$$C_r(f;x,x+h) - f(x) = O(h) \text{ as } h \to 0.$$

Theorem 1.8.9 *A function f is C_r-bounded at x if and only if $\underline{C_{r-1}Df}(x)$ and $\overline{C_rD}f(x)$ are finite.*

☐   This follows from the definition.   ☐

Theorem 1.8.10 *If f is C_r-bounded at x, then f is C_{r+1}-bounded at x.*

☐   By Theorem 1.8.9, $\underline{C_rD}f(x)$ and $\overline{C_rD}f(x)$ are finite and, hence, by Theorem 1.8.3, $\underline{C_{r+1}D}f(x)$ and $\overline{C_{r+1}D}f(x)$ are finite, and so by Theorem 1.8.9, f is C_{r+1}-bounded at x.   ☐

We now exhibit a function that will show that in Theorems 8.3 and 8.4 strict inequality is possible and will complete the discussion in Corollary 1.8.5.

Example. Let

$$f(x) = \begin{cases} 2x\sin\dfrac{1}{x^2} - \dfrac{2}{x}\cos\dfrac{1}{x^2}, & \text{if } x \neq 0, \\ 0, & \text{if } x = 0. \end{cases}$$

Then $C_0 Df(0) = -\infty, \overline{C_0 D}f(0) = \infty$, and remember that these derivates are the ordinary first order lower and upper derivates of f at 0. Since f is the first order derivative of F, where

$$F(x) = \begin{cases} x^2 \sin \dfrac{1}{x^2}, & \text{if } x \neq 0, \\ 0, & \text{if } x = 0, \end{cases}$$

f is special Denjoy integrable in every neighbourhood of 0 and F is its indefinite integral. So,

$$C_1(f; 0, h) = \frac{1}{h} \int_0^h f = \frac{F(h)}{h} = h \sin \frac{1}{h^2},$$

and, hence, f is C_1-continuous at 0 and

$$\underline{C_1 D}f(0) = \liminf_{h \to 0} \frac{2}{h}[C_1(f; 0, h) - f(0)] = \liminf_{h \to 0} 2 \sin \frac{1}{h^2} = -2.$$

Similarly, $\overline{C_1 D}f(0) = 2$. Also,

$$\begin{aligned} C_2(f; 0, h) &= \frac{2}{h^2} \int_0^h (h - t) f(t)\, dt = \frac{2}{h^2} \left[(h - t)F(t) \Big|_{t=0}^h + \int_0^h F \right] \\ &= \frac{2}{h^2} \int_0^h F = \frac{2}{h^2} \int_0^h t^2 \sin \frac{1}{t^2}\, dt, \end{aligned}$$

and, hence,

$$C_2 Df(0) = \lim_{h \to 0} \frac{3}{h}[C_2(f; 0, h) - f(0)] = \lim_{h \to 0} \frac{3!}{h^3} \int_0^h t^2 \sin \frac{1}{t^2}\, dt. \qquad (1.8.6)$$

Now

$$\frac{d}{dt}\left(t^5 \cos \frac{1}{t^2} \right) = 5t^4 \cos \frac{1}{t^2} + 2t^2 \sin \frac{1}{t^2}$$

and so

$$\int_0^h t^2 \sin \frac{1}{t^2}\, dt = \frac{1}{2} h^5 \cos \frac{1}{h^2} - \frac{5}{2} \int_0^h t^4 \cos \frac{1}{t^2}\, dt = o(h^4) \text{ as } h \to 0. \qquad (1.8.7)$$

From (1.8.6) and (1.8.7), $C_2 Df(0) = 0$. This shows that in Theorem 1.8.4 (ii) the inequalities may be strict and that for the last part of Corollary 1.8.5 the converse is not true.

Example. Consider further the function

$$g(x) = \begin{cases} 2 \sin \dfrac{1}{x^2} - \dfrac{2}{x^2} \cos \dfrac{1}{x^2} - \dfrac{4}{x^4} \sin \dfrac{1}{x^2}, & \text{if } x \neq 0, \\ 0, & \text{if } x = 0. \end{cases}$$

Then, with f as in the previous example, $f'(x) = g(x)$ for all $x \neq 0$. Since f is continuous everywhere except $x = 0$ where it is C_1-continuous and $\underline{C_1 D f}(0)$ and $\overline{C_1 D f}(0)$ are finite, f is both a $C_1 P$-major and a $C_1 P$-minor function of g and, hence, g is $C_1 P$-integrable and f is its indefinite $C_1 P$-integral. Integrating by parts,

$$
\begin{aligned}
C_2(g; 0, h) &= \frac{2}{h^2} \int_0^h (h-t) g(t) \, \mathrm{d}t = \frac{2}{h^2} \left[(h-t) f(t) \Big|_0^h + \int_0^h f \right] \\
&= \frac{2}{h^2} F(h) = 2 \sin \frac{1}{h^2}.
\end{aligned}
$$

Hence,

$$
\overline{C_2 D} g(0) = \limsup_{h \to 0} \frac{3}{h} [C_2(g; 0, h) - g(0)] = \limsup_{h \to 0} \frac{3!}{h} \sin \frac{1}{h^2} = \infty.
$$

Similarly, $\underline{C_2 D} g(0) = -\infty$. Since g is $C_1 P$-integrable, it is $C_2 P$-integrable and f is its indefinite $C_2 P$-integral. So,

$$
\begin{aligned}
C_3(g; 0, h) &= \frac{3}{h^3} \int_0^h (h-t)^2 g(t) \, \mathrm{d}t = \frac{3}{h^3} \left[(h-t)^2 f(t) \Big|_0^h + 2 \int_0^h (h-t) f(t) \, \mathrm{d}t \right] \\
&= \frac{3!}{h^3} \left[(h-t) F(t) \Big|_0^h + \int_0^h F \right] = \frac{3!}{h^3} \int_0^h t^2 \sin \frac{1}{t^2} \, \mathrm{d}t,
\end{aligned}
$$

and so by (1.8.7)

$$
C_3 D(g; 0, h) = \lim_{h \to 0} \frac{4}{h} [[C_3(g; 0, h) - g(0)] = \lim_{h \to 0} \frac{4!}{h^4} \int_0^h t^2 \sin \frac{1}{t^2} \, \mathrm{d}t = 0.
$$

There is no difficulty in considering one-sided limits and one-sided derivates and so the inequalities in Theorems 1.8.3 and 1.8.4 may be strict inequalities.

1.9　Symmetric Cesàro Derivatives

As for the Cesàro derivative, the symmetric Cesàro derivative of order r is defined with the help of the Cesàro–Perron integral of order $r - 1$. Let f be $C_{r-1} P$-integrable in $[a, b]$ and let the Cesàro mean $C_r(f; x, x + h)$ of f of order r on the interval with endpoints $x, x + h \in [a, b]$ be defined by (1.8.1).

1.9.1　Symmetric Cesàro Continuity and Symmetric Cesàro Derivative

Let

$$
\overline{SC}_r^+ f(x) = \limsup_{h \to 0+} [C_r(f; x, x + h) - C_r(f; x, x - h)],
$$

the other limits $\overline{SC_r^-}f(x), \underline{SC_r^+}f(x)$ and $\underline{SC_r^-}f(x)$ being defined analogously by using the appropriate limits of $[C_r(f;x,x+h) - C_r(f;x,x-h)]$. If $\overline{SC_r^+}f(x) = \overline{SC_r^-}f(x) = \underline{SC_r^+}f(x) = \underline{SC_r^-}f(x)$, then the common value, written $SC_r f(x)$, is called the *symmetric C_r-limit*, briefly the *SC_r-limit* of f at x. If $SC_r f(x) = 0$, then f is said to be *symmetrically C_r-continuous*, or *SC_r-continuous*, at x.

The *upper* and *lower SC_r derivates* of f at x are defined by

$$\overline{SC_rD}f(x) = \limsup_{h\to 0} \frac{r+1}{2h}[C_r(f;x,x+h) - C_r(f;x,x-h)],$$

$$\underline{SC_rD}f(x) = \liminf_{h\to 0} \frac{r+1}{2h}[C_r(f;x,x+h) - C_r(f;x,x-h)],$$

respectively. If $\overline{SC_rD}f(x) = \underline{SC_rD}f(x)$, the common value, denoted by $SC_rDf(x)$, is called the *SC_r-derivative* of f at x.

Note that for the concept of SC_rD there is no effect whether $h \to 0+$ or $h \to 0-$, but that it matters in the case of SC_r above.

Theorem 1.9.1 *If $r \geq 1$ and f is $C_{r-1}P$-integrable in $[a,b]$, then for $a < x < b$*

(i) $\underline{SC_r^+}f(x) \leq \underline{SC_{r+1}^+}f(x) \leq \overline{SC_{r+1}^+}f(x) \leq \overline{SC_r^+}f(x)$;

(ii) $\underline{SC_rD}f(x) \leq \underline{SC_{r+1}D}f(x) \leq \overline{SC_{r+1}D}f(x) \leq \overline{SC_rD}f(x)$.

☐   We prove (ii), the proof of (i) being similar and simpler. We may suppose that $\overline{SC_rD}f(x) < \infty$ and that $x = 0$.

Choose an arbitrary λ such that $\overline{SC_rD}f(x) < \lambda < \infty$. Then there is a $\delta > 0$ such that

$$C_r(f;0,\xi) - C_r(f;0,-\xi) < \lambda\frac{2\xi}{r+1} \text{ for } 0 < \xi < \delta.$$

So,

$$\frac{r}{\xi^r}\int_0^\xi (\xi - t)^{r-1}f(t)\, dt - \frac{r}{(-\xi)^r}\int_0^{-\xi}(-\xi - t)^{r-1}f(t)\, dt < \lambda\frac{2\xi}{r+1},$$

which gives

$$r\left[\int_0^\xi (\xi - t)^{r-1}f(t)\, dt + \int_0^{-\xi}(\xi + t)^{r-1}f(t)\, dt\right] < \lambda\frac{2\xi^{r+1}}{r+1} \text{ for } 0 < \xi < \delta.$$

Integrating,

$$r\left[\int_0^h \int_0^\xi (\xi - t)^{r-1}f(t)\, dt\, d\xi + \int_0^h \int_0^{-\xi}(\xi + t)^{r-1}f(t)\, dt\, d\xi\right]$$

$$\leq \lambda\frac{2h^{r+2}}{(r+1)(r+2)}, \text{ for } 0 < h < \delta. \qquad (1.9.1)$$

By Lemma 1.8.2

$$\int_0^h (h-t)^r f(t)\, dt = r \int_0^h \int_0^\xi (\xi - t)^{r-1} f(t)\, dt\, d\xi. \tag{1.9.2}$$

Since Lemma 1.8.2 is also true for negative values of h,

$$\int_0^{-h} (-h-t)^r f(t)\, dt = r \int_0^{-h} \int_0^\xi (\xi - t)^{r-1} f(t)\, dt\, d\xi$$

$$= -r \int_0^h \int_0^{-\xi} (-\xi - t)^{r-1} f(t)\, dt\, d\xi$$

and, hence,

$$\int_0^{-h} (h+t)^r f(t)\, dt = r \int_0^h \int_0^{-\xi} (\xi + t)^{r-1} f(t)\, dt\, d\xi. \tag{1.9.3}$$

From (1.9.1), (1.9.2) and (1.9.3)

$$\int_0^h (h-t)^r f(t)\, dt + \int_0^{-h} (h+t)^r f(t)\, dt \le \lambda \frac{2h^{r+2}}{(r+1)(r+2)}, \quad \text{for } 0 < h < \delta.$$

Hence,

$$\frac{r+2}{2h}\left[\frac{r+1}{h^{r+1}} \int_0^h (h-t)^r f(t)\, dt - \frac{r+1}{(-h)^{r+1}} \int_0^{-h} (-h-t)^r f(t)\, dt \right] \le \lambda,$$
$$\text{for } 0 < h < \delta,$$

which gives

$$\frac{r+2}{2h}\left[C_{r+1}(f; 0, h) - C_{r+1}(f; 0, -h) \right] \le \lambda, \text{ for } 0 < h < \delta.$$

Hence, $\overline{SC_{r+1}D}f(0) \le \lambda$. Since λ is arbitrary, $\overline{SC_{r+1}D}f(0) \le \overline{SC_r D}f(0)$, proving the last inequality of (ii).

The proof of the first inequality of (ii) is similar. □

Theorem 1.9.2 *If $r \ge 1$ and f is $C_{r-1}P$-integrable in $[a,b]$, then for $a < x < b$*

(i) *if f is SC_r-continuous at x, then f is SC_{r+1}-continuous at x;*

(ii) *if $SC_r Df(x)$ exists, then $SC_{r+1}Df(x)$ exists and $SC_r Df(x) = SC_{r+1}Df(x)$.*

□   Since $\overline{SC_r^-} f(x) = -\underline{SC_r^+} f(x)$, with three other similar relations, we get from (i) of Theorem 1.9.1

$$\underline{SC_r^-} f(x) \le \underline{SC_{r+1}^-} f(x) \le \overline{SC_{r+1}^-} f(x) \le \overline{SC_r^-} f(x). \tag{1.9.4}$$

So (i) follows from Theorem 1.9.1(i) and (1.9.4).

Also (ii) follows from Theorem 1.9.1(ii). □

Theorem 1.9.3 *If f is $C_0 P$-integrable in $[a, b]$, then for $a < x < b$*

(i) $\underline{SC_0^+} f(x) \leq \underline{SC_1^+} f(x) \leq \overline{SC}_1^+ f(x) \leq \overline{SC}_0^+ f(x)$;

(ii) $\underline{SC_0 D} f(x) \leq \underline{SC_1 D} f(x) \leq \overline{SC_1 D} f(x) \leq \overline{SC_0 D} f(x)$.

$\square$ As in Theorem 9.1, we prove (ii). Suppose that $x = 0$ and $\overline{SC_0 D} f(0) < \infty$. Choose λ, $\overline{SC_0 D} f(0) < \lambda < \infty$. Since $C_0(f; 0, t) = f(t)$ there is a $\delta > 0$ such that $f(t) - f(-t) < 2\lambda t$ for $0 < t < \delta$. Integrating we have

$$\frac{1}{h} \left[C_1(f; 0, h) - C_1(f; 0, -h) \right] = \frac{2}{2h} \left[\frac{1}{h} \int_0^h f - \frac{1}{-h} \int_0^{-h} f \right]$$

$$= \frac{2}{2h} \left[\frac{1}{h} \int_0^h f(t) \, dt - \frac{1}{h} \int_0^h f(-t) \, dt \right]$$

$$\leq \lambda \text{ for } 0 < h < \delta,$$

and so $\overline{SC_1 D} f(0) \leq \lambda$. Since λ is arbitrary, the right-hand inequality follows. The left-hand inequality has a similar proof. $\square$

Corollary 1.9.4 *If f is $C_0 P$-integrable in $[a, b]$, then for $a < x < b$*

(i) *if f is SC_0-continuous at x, then f is SC_1-continuous at x;*

(ii) *if $SC_0 D f(x)$ exists, then $SC_1 D f(x)$ exists and $SC_0 D f(x) = SC_1 D f(x)$.*

Remark. The SC_r-continuity of f at x is analogous to the smoothness of order $r + 1$ at x of the function that is obtained by taking the r times repeated integral of f. This will be discussed in Section 5 of Chapter II.

1.9.2   Symmetric Cesàro Boundedness

A function f is said to be *symmetric Cesàro bounded of order r at x*, or simply SC_r-*bounded at x* if

$$C_r(f; x, x + h) - C_r(f; x, x - h) = O(h) \text{ as } h \to 0.$$

Theorem 1.9.5 *A function f is SC_r-bounded at x if and only if $\overline{SC_r D} f(x)$ and $\underline{SC_r D} f(x)$ are finite. If f is SC_r-bounded at x, then f is SC_{r+1}-bounded at x.*

$\square$ The first part is clear and the second part follows from Theorem 1.9.1 and the first part of the theorem. $\square$

1.10 Borel Derivatives

1.10.1 Bilateral Borel Derivatives

Let f be special Denjoy integrable and let r be a fixed positive integer. If there is a polynomial $Q(t) = Q_x(t)$ of degree at most r such that

$$\int_0^h \frac{f(x+t) - Q(t)}{t^r}\, dt = o(h) \text{ as } h \to 0, \qquad (1.10.1)$$

then f is said to have a *Borel derivative of order r at x*, and if $a_r/r!$ is the coefficient of t^r in $Q(t)$, then a_r is called the *Borel derivative of f at x of order r* and is denoted by $BD_r f(x)$. The integral in (1.10.1) is assumed to be convergent at $t = 0$. Integrating by parts we have

$$
\begin{aligned}
\int_0^h \left(f(x+t) - Q(t)\right) dt &= \int_0^h t^r \frac{f(x+t) - Q(t)}{t^r}\, dt \\
&= h^r \int_0^h \frac{f(x+t) - Q(t)}{t^r}\, dt \\
&\quad - r \int_0^h t^{r-1} \int_0^t \frac{f(x+\xi) - Q(\xi)}{\xi^r}\, d\xi\, dt \\
&= o(h^{r+1}) \text{ as } h \to 0. \qquad (1.10.2)
\end{aligned}
$$

This, as we now see, shows that the Borel derivative, if it exists, is unique.

$\square$ If $Q_1(t)$ is another polynomial of degree at most r satisfying (1.10.1), then it also satisfies (1.10.2)and so

$$\int_0^h (Q - Q_1) = \int_0^h \left(f(x+t) - Q(t)\right) dt - \int_0^h \left(f(x+t) - Q_1(t)\right) dt = o(h^{r+1})$$

$$\text{as } h \to 0.$$

Since $\int_0^h (Q - Q_1)$ is a polynomial in h of degree at most $r + 1$, it vanishes identically and so $Q(t) = Q_1(t)$ for all t. $\square$

Now we show that if $r \geq 2$ and if $BD_r f(x)$ exists, then $BD_i f(x)$ exists, $0 < i < r$.

$\square$ For if $a_j/j!$ is the coefficient of t^j in $Q(t)$, $0 \leq j \leq r$, and if $\overline{Q}(t) = Q(t) - \sum_{j=i+1}^r (a_j/j!)t^j$, then using (1.10.1) and integrating by parts

$$
\begin{aligned}
\int_0^h \frac{f(x+t) - \overline{Q}(t)}{t^i}\, dt &= \int_0^h t^{r-i} \frac{f(x+t) - Q(t)}{t^r}\, dt + \int_0^h \sum_{j=i+1}^r \frac{a_j}{j!} t^{j-i}\, dt \\
&= h^{r-i} \int_0^h \frac{f(x+t) - Q(t)}{t^r}\, dt
\end{aligned}
$$

$$-(r-i)\int_0^h t^{r-i-1}\int_0^t \frac{f(x+\xi)-Q(\xi)}{\xi^r}\,d\xi\,dt+o(h)$$

$$=o(h^{r-i+1})+o(h)=o(h)\text{ as }h\to 0.$$

So (1.10.1) is satisfied if $Q(t)$ and r are replaced by $\overline{Q}(t)$ and i, respectively. Since $\overline{Q}(t)$ is a polynomial of degree at most i, $BD_i f(x)$ exists and $BD_i f(x)=a_i$. $\qquad\square$

If a_0 is the constant term in $Q(t)$, then writing $BD_0 f(x)=a_0$ we have from (1.10.2)

$$\int_0^h\left[f(x+t)-\sum_{i=0}^r \frac{t^i}{i}BD_i f(x)\right]dt=o(h^{r+1})\text{ as }h\to 0.$$

and so

$$BD_0 f(x)=\lim_{h\to 0}\frac{1}{h}\int_0^h f(x+t)\,dt,$$

which is $f(x)$ if f is continuous at x. So we make the following definition.

Definition. If $\lim_{h\to 0}\frac{1}{h}\int_0^h f(x+t)\,dt=f(x)$, then f is said to be *Borel continuous* at x.

Thus, Borel continuity is the same as Cesàro continuity of order 1, C_1-continuity. We shall see in Chapter II that a Borel derivative is not the same as a Cesàro derivative. To define upper and lower Borel derivates of order k at x, we suppose the existence of $BD_{k-1}f(x)$ and define

$$Q(t)=\sum_{i=0}^{k-1}\frac{t^i}{i!}BD_i f(x). \tag{1.10.3}$$

Then for each $h>0$ and $0<\epsilon<h$, $1/t$ is of bounded variation in $[\epsilon,h]$ and so the function $k!\left[f(x+t)-Q(t)\right]/t^k$ is special Denjoy integrable in $[\epsilon,h]$ as a function of t. The *right upper and lower Borel derivates of f at x of order k* are defined by

$$\overline{BD}_k^+ f(x)=k!\limsup_{h\to 0+}\left[\limsup_{\epsilon\to 0+}\frac{1}{h}\int_\epsilon^h \frac{f(x+t)-Q(t)}{t^k}\,dt\right],$$

$$\underline{BD}_k^+ f(x)=k!\liminf_{h\to 0+}\left[\liminf_{\epsilon\to 0+}\frac{1}{h}\int_\epsilon^h \frac{f(x+t)-Q(t)}{t^k}\,dt\right], \tag{1.10.4}$$

respectively. If $\overline{BD}_k^+ f(x)=\underline{BD}_k^+ f(x)$, the common value, possibly infinite, is called the *right Borel derivative of f at x of order k* and is denoted by $BD_k^+ f(x)$. The definitions of $\overline{BD}_k^- f(x),\underline{BD}_k^- f(x)$ and $BD_k^- f(x)$ are obtained by considering $[-h,-\epsilon]$ and taking the integral from $-h$ to $-\epsilon$, where $h>0$ and $0<\epsilon<h$. If $\overline{BD}_k^+ f(x)=\underline{BD}_k^+ f(x)=\overline{BD}_k^- f(x)=\underline{BD}_k^- f(x)$, then

the common value, possibly infinite, is the Borel derivative of f at x of order k, written $BD_k f(x)$. We now show that if $BD_k f(x)$ exists finitely, then this definition agrees with the previous one.

☐   Writing

$$F(x, \epsilon, h) = \int_\epsilon^h \frac{f(x+t) - Q(t)}{t^k}\, dt, \tag{1.10.5}$$

where Q is defined in (1.10.3), observe that if $\overline{BD}_k^+ f(x)$ and $\underline{BD}_k^+ f(x)$ are finite, then $\limsup_{\epsilon \to 0+} F(x, \epsilon, h)$ and $\liminf_{\epsilon \to 0+} F(x, \epsilon, h)$ are finite. We first show that they are equal. Let

$$\Phi(x, h) = \limsup_{\epsilon \to 0+} F(x, \epsilon, h),$$

$$\phi(x, h) = \liminf_{\epsilon \to 0+} F(x, \epsilon, h). \tag{1.10.6}$$

Since

$$\overline{BD}_k^+ f(x) = \limsup_{h \to 0+} \frac{\Phi(x, h)}{h} \quad \text{and} \quad \underline{BD}_k^+ f(x) = \liminf_{h \to 0+} \frac{\phi(x, h)}{h},$$

and since they are finite, we have

$$\limsup_{h \to 0+} \Phi(x, h) = 0 = \liminf_{h \to 0+} \phi(x, h),$$

and, therefore, since $\phi(x, h) \le \Phi(x, h)$,

$$\lim_{h \to 0+} \big(\Phi(x, h) - \phi(x, h)\big) = 0. \tag{1.10.7}$$

Let $\eta > 0$ be small, then

$$\Phi(x, h + \eta) - \Phi(x, h) \tag{1.10.8}$$

$$= \limsup_{\epsilon \to 0+} \int_\epsilon^{h+\eta} \frac{f(x+t) - Q(t)}{t^k}\, dt - \limsup_{\epsilon \to 0+} \int_\epsilon^h \frac{f(x+t) - Q(t)}{t^k}\, dt$$

$$= \limsup_{\epsilon \to 0+} \left[\int_\epsilon^h + \int_h^{h+\eta} \right] \frac{f(x+t) - Q(t)}{t^k}\, dt - \limsup_{\epsilon \to 0+} \int_\epsilon^h \frac{f(x+t) - Q(t)}{t^k}\, dt$$

$$= \int_h^{h+\eta} \frac{f(x+t) - Q(t)}{t^k}\, dt.$$

Similarly,

$$\phi(x, h + \eta) - \phi(x, h) = \int_h^{h+\eta} \frac{f(x+t) - Q(t)}{t^k}\, dt. \tag{1.10.9}$$

From (1.10.8) and (1.10.9),

$$\Phi(x, h + \eta) - \Phi(x, h) = \phi(x, h + \eta) - \phi(x, h)$$

or
$$\Phi(x, h + \eta) - \phi(x, h + \eta) = \Phi(x, h) - \phi(x, h).$$

This shows that $\Phi(x, h) - \phi(x, h)$ is independent of h and so from (1.10.7), $\Phi(x, h) = \phi(x, h)$ for all h. Hence, from (1.10.5) and (1.10.6), $\lim_{\epsilon \to 0} \int_{\epsilon}^{h} \frac{f(x+t)-Q(t)}{t^k} dt$ exists and so from (1.10.4), if $\overline{BD}_k^+ f(x)$ and $\underline{BD}_k^+ f(x)$ are finite, then

$$\overline{BD}_k^+ f(x) = k! \limsup_{h \to 0+} \frac{1}{h} \int_0^h \frac{f(x+t) - Q(t)}{t^k} dt,$$

$$\underline{BD}_k^+ f(x) = k! \liminf_{h \to 0+} \frac{1}{h} \int_0^h \frac{f(x+t) - Q(t)}{t^k} dt. \qquad (1.10.10)$$

Now suppose that $BD_k f(x)$ exists finitely in this latter sense. Then from (1.10.10) and its analogue

$$\frac{1}{h} \int_0^h \frac{f(x+t) - Q(t)}{t^k} dt = \frac{1}{k!} BD_k f(x) + o(1) \text{ as } h \to 0,$$

and so

$$\frac{1}{h} \int_0^h \frac{f(x+t) - Q(t) - \frac{t^k}{k!} BD_k f(x)}{t^k} dt = o(1) \text{ as } h \to 0,$$

and this gives by (1.10.3)

$$\frac{1}{h} \int_0^h \frac{f(x+t) - Q(t) - \sum_{i=0}^k \frac{t^i}{i!} BD_k f(x)}{t^k} dt = o(1) \text{ as } h \to 0,$$

showing that $BD_k f(x)$ is also the Borel derivative in the former sense. □

1.10.2 Unilateral Borel Derivatives

If in (1.10.1) a one-sided limit is taken, the definition of the one-sided Borel derivatives $BD_r^+ f(x)$ and $BD_r^- f(x)$ are obtained and if in (1.10.3) we take

$$Q(t) = \sum_{i=0}^{k-1} \frac{t^i}{i!} BD_i^+ f(x), \qquad (1.10.3^*)$$

the definition of the unilateral Borel derivates are obtained.

Note that the bilateral Borel derivates and the unilateral Borel derivates are different since (1.10.3)* and (1.10.3) are different.

1.10.3 Borel Boundedness

If in (1.10.1) the condition $o(1)$ is replaced by $O(1)$ as $t \to 0$, then f is said to be *Borel bounded of order r at x*.

Theorem 1.10.1 *If f is Borel bounded of order r at x, then f has a finite Borel derivative of order $r - 1$ and $\overline{BD}_r f(x)$ and $\underline{BD}_r f(x)$ are finite. Conversely, if $\overline{BD}_r f(x)$ and $\underline{BD}_r f(x)$ are finite, then f is Borel bounded of order r at x.*

☐   We have from the definition of Borel boundedness that

$$\frac{1}{h} \int_0^h \frac{f(x+t) - Q(t)}{t^r}\, dt = O(1) \text{ as } h \to 0, \tag{1.10.11}$$

where $Q(t)$ is a polynomial of degree at most r. Let $Q(t) = \sum_{i=0}^r b_i t^i$. Writing $\overline{Q}(t) = \sum_{i=0}^{r-1} b_i t^i$, we have from (1.10.11)

$$
\begin{aligned}
\int_0^h \frac{f(x+t) - \overline{Q}(t)}{t^{r-1}}\, dt &= \int_0^h t\frac{f(x+t) - Q(t)}{t^r}\, dt + \int_0^h b_r t\, dt \\
&= h\int_0^h \frac{f(x+t) - Q(t)}{t^r}\, dt \\
&\quad - \int_0^h \int_0^t \frac{f(x+\xi) - Q(\xi)}{\xi^r}\, d\xi\, dt + O(h^2) \\
&= hO(h) - \int_0^h O(t)\, dt + O(h^2) = O(h^2) \text{ as } h \to 0.
\end{aligned}
$$

Hence,

$$\frac{1}{h} \int_0^h \frac{f(x+t) - \overline{Q}(t)}{t^{r-1}}\, dt = O(h) \text{ as } h \to 0. \tag{1.10.12}$$

Since $O(h)$ implies $o(h)$ as $h \to 0$, (1.10.12) shows that $BD_{r-1}f(x)$ exists finitely and that $\overline{Q}(t) = \sum_{i=0}^{r-1}(t^i/i!)BD_i f(x)$. Since $Q(t) = \overline{Q}(t) + b_r t^r$, from (1.10.11),

$$
\begin{aligned}
\frac{1}{h} \int_0^h \frac{f(x+t) - \sum_{i=0}^{r-1}(t^i/i!)BD_i f(x)}{t^r}\, dt \\
= \frac{1}{h}\int_0^h b_r\, dt + O(1) = O(1) \text{ as } h \to 0. \tag{1.10.13}
\end{aligned}
$$

Hence, $\overline{BD}_r f(x)$ and $\underline{BD}_r f(x)$ are finite.

For the converse, since $\overline{BD}_r f(x)$ and $\underline{BD}_r f(x)$ are finite by definition, (1.10.13) holds. Let C be any finite constant and let $Q(t) = \sum_{i=0}^{r-1}(t^i/i!)BD_i f(x) + Ct^r$. Then from (1.10.13)

$$\frac{1}{h} \int_0^h \frac{f(x+t) - Q(t)}{t^r}\, dt = O(1) - \frac{1}{h}\int_0^h C\, dt = O(1) \text{ as } h \to 0.$$

Since $Q(t)$ is a polynomial of degree of most r, f is Borel bounded of order r at x.   ☐

1.11 Symmetric Borel Derivatives

1.11.1 Symmetric Borel Derivatives and Symmetric Borel Continuity

Let f be special Denjoy integrable in some neighbourhood of x and let r be a positive integer. If there is a polynomial $P(t) = P_x(t)$ of degree at most r such that

$$\int_0^h \frac{\frac{1}{2}(f(x+t) + (-1)^r f(x-t)) - P(t)}{t^r}\, dt = o(h), \text{ as } h \to 0, \qquad (1.11.1)$$

then f is said to have a *symmetric Borel derivative* at x of order r and if $a_r/r!$ is the coefficient of t^r in $P(t)$, then a_r is called the *symmetric Borel derivative* of f at x of order r, and is denoted by $SBD_r f(x)$. The integral in (1.11.1) is assumed to be convergent at $t = 0$.

The derivative, if it exists, is unique.

☐ Integrating by parts and using (1.11.1), we have

$$\int_0^h \left[\tfrac{1}{2}(f(x+t) + (-1)^r f(x-t)) - P(t) \right] dt$$

$$= \int_0^h t^r \frac{\frac{1}{2}(f(x+t) + (-1)^r f(x-t)) - P(t)}{t^r}\, dt$$

$$= h^r \int_0^h \frac{\frac{1}{2}(f(x+t) + (-1)^r f(x-t)) - P(t)}{t^r}\, dt$$

$$- r \int_0^h t^{r-1} \int_0^t \frac{\frac{1}{2}(f(x+\xi) + (-1)^r f(x-\xi)) - P(\xi)}{\xi^r}\, d\xi\, dt$$

$$= o(h^{r+1}), \text{ as } h \to 0. \qquad (1.11.2)$$

So, if $P_1(t)$ is another polynomial of degree at most r satisfying (1.11.1), then it also satisfies (1.11.2) and so

$$\int_0^h \left[P_1(t) - P(t) \right] dt = \int_0^h \left[\tfrac{1}{2}(f(x+t) + (-1)^r f(x-t)) - P(t) \right] dt$$

$$- \int_0^h \left[\tfrac{1}{2}(f(x+t) + (-1)^r f(x-t)) - P_1(t) \right] dt$$

$$= o(h^{r+1}), \text{ as } h \to 0. \qquad (1.11.3)$$

Since $P_1(t) - P(t)$ is a polynomial of degree at most r, the left-hand side of (11.3) is a polynomial in h of degree at most $r + 1$ and so from (1.11.3), $P_1(t) - P(t)$ vanishes identically. Hence, $P_1(t) = P(t)$ for all t. Therefore, if $SBD_r f(x)$ exists, it is unique. ☐

We now show that the polynomial $P(t)$ in (1.11.1) has only even or only odd powers of t according as r is even or odd.

□   For, since (1.11.2) is true for all h, changing h to $-h$ in (1.11.2) we have

$$\int_{-h}^{0} \left[\tfrac{1}{2}(f(x+t) + (-1)^r f(x-t)) - P(t) \right] dt = o(h^{r+1}) \text{ as } h \to 0.$$

Again, changing the variable t to $-t$ and multiplying by $(-1)^r$

$$\int_{0}^{h} \left[\tfrac{1}{2}(f(x+t) + (-1)^r f(x-t)) - (-1)^r P(-t) \right] dt = o(h^{r+1}) \text{ as } h \to 0.$$

(1.11.4)

From (1.11.2) and (1.11.4),

$$\int_{0}^{h} \left(P(t) - (-1)^r P(-t) \right) dt = o(h^{r+1}) \text{ as } h \to 0. \tag{1.11.5}$$

Since $P(t) - (-1)^r P(-t)$ is a polynomial of degree at most r, the left-hand side of (1.11.5) is a polynomial in h of degree at most $r+1$ and so $P(t) - (-1)^r P(-t)$ vanishes identically. Hence, $P(t)$ has only even or odd powers of t according as r is even or odd.   □

If $r \geq 2$ and $SBD_r f(x)$ exists, then $SBD_{r-2} f(x)$ exists.

□   For if $P(t) = \sum_{i=0}^{r} a_i t^i$, then writing $S(t) = P(t) - a_r t^r$ and, since $P(t)$ has only odd or even powers, we conclude that $S(t)$ is of degree at most $r - 2$. Also from (1.11.1) integration by parts gives

$$\int_{0}^{h} \frac{\tfrac{1}{2}(f(x+t) + (-1)^r f(x-t)) - S(t)}{t^{r-2}} \, dt$$

$$= \int_{0}^{h} \frac{\tfrac{1}{2}(f(x+t) + (-1)^r f(x-t)) - P(t) + a_r t^r}{t^{r-2}} \, dt$$

$$= \int_{0}^{h} t^2 \frac{\tfrac{1}{2}(f(x+t) + (-1)^r f(x-t)) - P(t)}{t^r} \, dt + a_r \frac{h^3}{3}$$

$$= h^2 \int_{0}^{h} \frac{\tfrac{1}{2}(f(x+t) + (-1)^r f(x-t)) - P(t)}{t^r} \, dt$$

$$\quad -2 \int_{0}^{h} t \int_{0}^{t} \frac{\tfrac{1}{2}(f(x+\xi) + (-1)^r f(x-\xi)) - P(\xi)}{\xi^r} \, d\xi \, dt + a_r \frac{h^3}{3}$$

$$= h^2 o(h) - 2 \int_{0}^{h} t \big(o(t) \big) \, dt + a_r \frac{h^3}{3} = o(h^2) \text{ as } h \to 0.$$

Since $o(h^2)$ implies $o(h)$, as $h \to 0$, $SBD_{r-2} f(x)$ exists.   □

When r is even, we have to consider $SBD_0 f(x)$, which is given by

$$SBD_0 f(x) = \lim_{h \to 0} \frac{1}{h} \int_{0}^{h} \tfrac{1}{2}(f(x+t) + f(x-t)) \, dt,$$

which is $f(x)$ if f is continuous at x. So, we make the following definitions.

Definitions. If

$$\lim_{h \to 0} \frac{1}{h} \int_0^h \tfrac{1}{2} \big(f(x+t) + f(x-t) \big) \, dt = f(x),$$

then f is said to be *symmetrically Borel continuous at x of even order*. If

$$\lim_{h \to 0} \frac{1}{h} \int_0^h \tfrac{1}{2} \big(f(x+t) - f(x-t) \big) \, dt = 0,$$

then f is said to be *symmetrically Borel continuous at x of odd order*.

Clearly symmetric Borel continuity of odd order is equivalent to SC_1-continuity. Also, it is clear that symmetric continuity of even, (odd) order (see Section 6.1 for the definitions), implies symmetric Borel continuity of even, (odd) order.

To define upper and lower symmetric Borel derivates of f at x of order k, $k \geq 2$, we suppose that $SBD_{k-2}f(x)$ exists and define $\overline{\omega}_k(f; x, t)$ as follows:

$$t\overline{\omega}_1(f; x, t) = \tfrac{1}{2} \big[f(x+t) - f(x-t) \big],$$

$$(1.11.6)$$

$$\frac{t^k}{k!} \overline{\omega}_k(f; x, t) = \tfrac{1}{2} \big[f(x+t) + (-1)^{k-2} f(x-t) \big] - P(t) \text{ for } k \geq 2,$$

where $P(t)$ is the polynomial of degree at most $k-2$ given by

$$P(t) = \begin{cases} \displaystyle\sum_{i=0}^{(k-2)/2} \frac{t^{2i}}{(2i)!} SBD_{2i}f(x), & \text{if } k \text{ is even,} \\[4mm] \displaystyle\sum_{i=0}^{(k-3)/2} \frac{t^{2i+1}}{(2i+1)!} SBD_{2i+1}f(x), & \text{if } k \text{ is odd.} \end{cases}$$

$$(1.11.7)$$

For small $h > 0$ and $0 < \epsilon < h$, the function $1/t$ is of bounded variation in $[\epsilon, h]$ and so $\overline{\omega}_1(f; x, t)$, as a function of t, is special Denjoy integrable in $[\epsilon, h]$. The *upper and lower symmetric Borel derivates of f at x of order k* are defined by

$$\overline{SBD}_k f(x) = \limsup_{h \to 0+} \limsup_{\epsilon \to 0+} \frac{1}{h} \int_\epsilon^h \overline{\omega}_k(f; x, t) \, dt,$$

$$\underline{SBD}_k f(x) = \liminf_{h \to 0+} \liminf_{\epsilon \to 0+} \frac{1}{h} \int_\epsilon^h \overline{\omega}_k(f; x, t) \, dt,$$

respectively. If $\overline{SBD}_k f(x) = \underline{SBD}_k f(x)$, then the common value, denoted by $SBD_k f(x)$, is called the *symmetric Borel derivative, possibly infinite*, of

f at x of order k. It can be shown, as in the case of the unsymmetric Borel derivative, that if $\overline{SBD}_k f(x)$ and $\underline{SBD}_k f(x)$ are finite, then

$$\overline{SBD}_k f(x) = \limsup_{h \to 0+} \frac{1}{h} \int_0^h \overline{\omega}_k(f; x, t)\, dt,$$

$$\underline{SBD}_k f(x) = \liminf_{h \to 0+} \frac{1}{h} \int_0^h \overline{\omega}_k(f; x, t)\, dt, \qquad (1.11.8)$$

and that if $SBD_k f(x)$ exists finitely in this latter sense, then it also exists in the former sense and they are equal.

Note that $h \to 0+$ in (1.11.8) can be replaced by $h \to 0$.

1.11.2   Borel Smoothness

Let a function f be defined in some neighbourhood of x and let f be special Denjoy integrable in that neighbourhood. Let $SBD_{k-2} f(x)$ exist finitely. Define

$$\overline{BS}_k f(x) = \limsup_{h \to 0} \frac{1}{h} \int_0^h t\overline{\omega}_k(f; x, t)\, dt,$$

$$\underline{BS}_k f(x) = \liminf_{h \to 0} \frac{1}{h} \int_0^h t\overline{\omega}_k(f; x, t)\, dt. \qquad (1.11.9)$$

Then $\overline{BS}_k f(x)$ and $\underline{BS}_k f(x)$ are, respectively, called the *index of upper and lower Borel smoothness* of f of order k at x. If $\overline{BS}_k f(x) = \underline{BS}_k f(x)$, then the common value, denoted by $BS_k f(x)$, is called the *index of Borel smoothness* of f of order k at x. If $BS_k f(x) = 0$, then f is said to be *Borel smooth* of order k at x.

Theorem 1.11.1 *If $\overline{SBD}_k f(x)$ and $\underline{SBD}_k f(x)$ are finite, then f is Borel smooth of order k at x.*

□   Since $\overline{SBD}_k f(x)$ and $\underline{SBD}_k f(x)$ are finite, it follows from (1.11.8) that $\lim_{h \to 0} \int_0^h \overline{\omega}_k(f; x, t)\, dt = 0$. So, integrating by parts,

$$\int_0^h t\overline{\omega}_k(f; x, t)\, dt = h \int_0^h \overline{\omega}_k(f; x.t)\, dt - \int_0^h \int_0^t \overline{\omega}_k(f; x, \xi)\, d\xi\, dt$$

and hence,

$$\frac{1}{h} \int_0^h t\overline{\omega}_k(f; x, t)\, dt = \int_0^h \overline{\omega}_k(f; x.t)\, dt - \frac{1}{h} \int_0^h \int_0^t \overline{\omega}_k(f; x, \xi)\, d\xi\, dt,$$

and so the left-hand side tends to 0 as $h \to 0$.   □

Corollary *If f is Borel smooth of order k at x, then f is Borel smooth of order $k-2$ at x.*

☐ If f is Borel smooth of order k at x, then by definition $SBD_{k-2}f(x)$ exists finitely and so the result follows from the above theorem. ☐

1.11.3 Symmetric Borel Boundedness

If, in (1.11.1), $o(h)$ is replaced by $O(h)$ as $h \to 0$, then f is said to be *symmetric Borel bounded of order r at x.*

Theorem 1.11.2 *If f is symmetric Borel bounded of order r at x, then f has a finite symmetric Borel derivative of order $r-2$ at x and $\overline{SBD}_r f(x)$ and $\underline{SBD}_r f(x)$ are finite. Conversely, if $\overline{SBD}_r f(x)$ and $\underline{SBD}_r f(x)$ are finite, then f is symmetric Borel bounded of order r at x.*

☐ We have from the definition of symmetric Borel boundedness

$$\frac{1}{h} \int_0^h \frac{\frac{1}{2}\big(f(x+t) + (-1)^r f(x-t)\big) - P(t)}{t^r} \, dt = O(1), \quad \text{as } h \to 0, \quad (1.11.12)$$

where $P(t)$ is a polynomial of degree at most r. Let $P(t) = \sum_{i=0}^r b_i t^i$. Then writing $\overline{P}(t) = \sum_{i=0}^{r-2} b_i t^i$ we have from (1.11.12)

$$\int_0^h \frac{\frac{1}{2}\big(f(x+t) + (-1)^{r-2} f(x-t)\big) - \overline{P}(t)}{t^{r-2}} \, dt$$

$$= \int_0^h t^2 \frac{\frac{1}{2}\big(f(x+t) + (-1)^r f(x-t)\big) - P(t)}{t^r} \, dt + \int_0^h (b_{r-1}t + b_r t^2) \, dt$$

$$= h^2 \int_0^h \frac{\frac{1}{2}\big(f(x+t) + (-1)^r f(x-t)\big) - P(t)}{t^r} \, dt$$

$$- 2 \int_0^h t \int_0^t \frac{\frac{1}{2}\big(f(x+\xi) + (-1)^r f(x-\xi)\big) - P(\xi)}{\xi^r} \, d\xi \, dt + O(h^2)$$

$$= h^2 O(h) - 2 \int_0^h t O(t) \, dt + O(h^2) = O(h^2) \text{ as } h \to 0.$$

Hence,

$$\frac{1}{h} \int_0^h \frac{\frac{1}{2}\big(f(x+t) + (-1)^{r-2} f(x-t)\big) - \overline{P}(t)}{t^{r-2}} \, dt = O(h) \text{ as } h \to 0. \quad (1.11.13)$$

Since $O(h)$ implies $o(1)$ as $h \to 0$ and $\overline{P}(t)$ is a polynomial of degree at most $r-2$, (1.11.13) shows that $SBD_{r-2}f(x)$ exists and that $SBD_{r-2}f(x) = (r-2)! b_{r-2}$.

We now show that $b_{r-1} = 0$. Using (1.11.12)

$$\int_0^h \left(\frac{1}{2}\big(f(x+t) + (-1)^r f(x-t)\big) - P(t)\right) dt$$

$$= \int_0^h t^r \frac{\frac{1}{2}\big(f(x+t) + (-1)^r f(x-t)\big) - P(t)}{t^r} \, dt$$

$$= h^r \int_0^h \frac{\frac{1}{2}\big(f(x+t) + (-1)^r f(x-t)\big) - P(t)}{t^r} \, dt \qquad (1.11.14)$$

$$- r \int_0^h t^{r-1} \int_0^t \frac{\frac{1}{2}\big(f(x+\xi) + (-1)^r f(x-\xi)\big) - P(\xi)}{\xi^r} \, d\xi \, dt$$

$$= h^r O(h) - r \int_0^h t^{r-1} O(t) \, dt = O(h^{r+1}) \text{ as } h \to 0.$$

Since (1.11.14) holds for all h,

$$\int_0^{-h} \left(\frac{1}{2}\big(f(x+t) + (-1)^r f(x-t)\big) - P(t)\right) dt = O(h^{r+1}) \text{ as } h \to 0.$$

Changing the variable t to $-t$ and multiplying by $-(-1)^r$, we get

$$\int_0^h \left(\frac{1}{2}\big(f(x+t) + (-1)^r f(x-t)\big) - (-1)^r P(-t)\right) dt = O(h^{r+1}) \text{ as } h \to 0.$$

$$(1.11.15)$$

Subtracting (1.11.14) from (1.11.15),

$$\int_0^h \big(P(t) - (-1)^r P(-t)\big) \, dt = O(h^{r+1}) \text{ as } h \to 0. \qquad (1.11.16)$$

Since $P(t) - (-1)^r P(-t)$ is a polynomial of degree at most r, then $\int_0^h (P(t) - (-1)^r P(-t)) \, dt$ is a polynomial in h of degree at most $r + 1$ and so from (1.11.16), $P(t) - (-1)^r P(-t) = Ct^r$ for some constant C. Hence, if r is even, $P(t)$ does not contain any odd powers of t and, if r is odd, then $P(t)$ does not contain any even powers of t. So, $P(t)$ does not contain the term $b_{r-1} t^{r-1}$; that is, $b_{r-1} = 0$ and $P(t) = \overline{P}(t) + b_r t^r$.

Let r be even, $r = 2m$, say. Then from (1.11.13), $\overline{P}(t) = \sum_{i=0}^{m-1} \frac{t^{2i}}{(2i)!} SBD_{2i} f(x)$ and so $P(t) = \sum_{i=0}^{m-1} \frac{t^{2i}}{(2i)!} SBD_{2i} f(x) + b_{2m} t^{2m}$ and, hence, from (1.11.12),

$$\frac{1}{h} \int_0^h \frac{\left(\frac{1}{2} f(x+t) + f(x-t) - \sum_{i=0}^{m-1} \frac{t^{2i}}{(2i)!} SBD_{2i} f(x)\right)}{t^{2m}} \, dt$$

$$= \frac{1}{h} \int_0^h b_{2m} \, dt + O(1) = O(1) \text{ as } h \to 0, \qquad (1.11.17)$$

which shows that $\overline{SBD}_{2m} f(x)$ and $\underline{SBD}_{2m} f(x)$ are finite.

If r is odd, $r = 2m + 1$, say, then from (1.11.13), $\overline{P}(t) = \sum_{i=1}^{m} \frac{t^{2i-1}}{(2i-1)!} SBD_{2i-1}f(x)$ and so $P(t) = \sum_{i=1}^{m} \frac{t^{2i-1}}{(2i-1)!} SBD_{2i-1}f(x) + b_{2m+1}t^{2m+1}$ and, hence, from (1.11.12)

$$\frac{1}{h} \int_0^h \frac{\left(\frac{1}{2}f(x+t) - f(x-t) - \sum_{i=1}^{m} \frac{t^{2i-1}}{(2i-1)!} SBD_{2i-1}f(x)\right)}{t^{2m+1}} \, dt$$

$$= \frac{1}{h} \int_0^h b_{2m+1} \, dt + O(1) = O(1) \text{ as } h \to 0, \tag{1.11.18}$$

which shows that $\overline{SBD}_{2m+1}f(x)$ and $\underline{SBD}_{2m+1}f(x)$ are finite.

For the converse part, since $\overline{SBD}_r f(x)$ and $\underline{SBD}_r f(x)$ are finite, by definition (1.11.17) or (1.11.18) holds according as $r = 2m$ or $r = 2m + 1$. Let C be any finite real constant and let

$$P(t) = \begin{cases} \displaystyle\sum_{i=0}^{m-1} \frac{t^{2i}}{(2i)!} SBD_{2i}f(x) + Ct^{2m}, & \text{if } r = 2m, \\[3mm] \displaystyle\sum_{i=1}^{m} \frac{t^{2i-1}}{(2i-1)!} SBD_{2i-1}f(x) + Ct^{2m+1}, & \text{if } r = 2m+1. \end{cases}$$

Then from (1.11.17) or (1.11.18)

$$\frac{1}{h} \int_0^h \frac{\frac{1}{2}\left(f(x+t) + (-1)^r f(x-t)\right) - P(t)}{t^r} \, dt = O(1) - \frac{1}{h} \int_0^h C \, dt = O(1),$$

as $h \to 0$, which shows that f is symmetric Borel bounded of order r at x. $\square$

1.12 L^p-Derivatives

1.12.1 L^p-Derivatives and L^p-Continuity

Let $f \in L^p$ in some neighbourhood of a point x for some p, $1 \le p < \infty$. If there is a polynomial $P(t) = P_x(t)$ of degree at most k for which

$$\left[\frac{1}{h} \int_0^h |f(x+t) - P(t)|^p \, dt\right]^{1/p} = o(h^k) \text{ as } h \to 0, \tag{1.12.1}$$

then f is said to have a kth L^p-derivative at x of order k and, if $a_k/k!$ is the coefficient of x^k in $P(t)$, then a_k is called the kth L^p-derivative of f at x denoted by $f_{(k),p}(x)$. If $f_{(k),p}(x)$ exists, then it is unique.

☐   Let $P_1(t)$ be another polynomial of degree at most k that satisfies (1.12.1). Then by Minkowski's inequality

$$\left[\frac{1}{h}\int_0^h |P_1(t) - P(t)|^p \, dt\right]^{1/p} \leq \left[\frac{1}{h}\int_0^h |f(x+t) - P(t)|^p \, dt\right]^{1/p}$$

$$+ \left[\frac{1}{h}\int_0^h |f(x+t) - P_1(t)|^p \, dt\right]^{1/p}$$

$$= o(h^k) \text{ as } h \to 0.$$

Taking $h > 0$ we have

$$\frac{1}{h}\int_0^h |P_1(t) - P(t)|^p \, dt = o(h^{kp}) \text{ as } h \to 0. \tag{1.12.2}$$

Let $Q(t) = P_1(t) - P(t) = \sum_{i=0}^k c_i t^i$. Since $|Q(t)|^p$ is a continuous function of t, we have from (1.12.2), $|Q(0)|^p = \lim_{h\to 0+} \frac{1}{h}\int_0^h |Q(t)|^p \, dt = 0$, and so $c_0 = 0$. Suppose that we have proved $c_0 = c_1 = \cdots = c_j = 0$, $0 \leq j < k$. Then $Q(t) = \sum_{i=j+1}^k c_i t^i$ and so integrating by parts,

$$\int_0^h |Q(t)|^p \, dt = \int_0^h t^{p(j+1)} \left|\sum_{i=j+1}^k c_i t^{i-j-1}\right|^p \, dt$$

$$= h^{p(j+1)} \int_0^h \left|\sum_{i=j+1}^k c_i t^{i-j-1}\right|^p \, dt$$

$$-p(j+1) \int_0^h t^{p(j+1)-1} \int_0^t \left|\sum_{i=j+1}^k c_i \xi^{i-j-1}\right|^p \, d\xi \, dt.$$

Hence,

$$\frac{1}{h^{pj+p+1}}\int_0^h |Q(t)|^p \, dt = \frac{1}{h}\int_0^h \left|\sum_{i=j+1}^k c_i t^{i-j-1}\right|^p \, dt$$

$$-\frac{p(j+1)}{h^{pj+p+1}}\int_0^h t^{p(j+1)-1} \int_0^t \left|\sum_{i=j+1}^k c_i \xi^{i-j-1}\right|^p \, d\xi \, dt.$$

$$\tag{1.12.3}$$

Now as $h \to 0$, the right-hand side of (1.12.3) tends to $c_{j+1} - \frac{pj+p}{pj+p+1}c_{j+1}$, that is, to $c_{j+1}/(pj+p+1)$. But, by (1.12.2), the left-hand side of (1.12.3) tends to 0 and so $c_{j+1} = 0$.

Hence, by induction, $c_j = 0$, $0 \leq j \leq k$ and so $P_1(t) = P(t)$. If $h < 0$, the proof is similar.   ☐

If $f_{(k),p}(x)$ exists and $k \geq 2$, then $f_{(k-1),p}(x)$ also exists.

☐   To see this, note that if $h > 0$, then

$$\left[\frac{1}{h}\int_0^h |t^k|^p \, dt\right]^{1/p} = \left[\frac{1}{h}\int_0^h t^{kp} \, dt\right]^{1/p} = \left[\frac{h^{kp}}{kp+1}\right]^{1/p} = O(h^k) \text{ as } h \to 0,$$

$$(1.12.4)$$

and, if $h < 0$, then

$$\left[\frac{1}{h}\int_0^h |t^k|^p \, dt\right]^{1/p} = \left[\frac{1}{-h}\int_h^0 |t^k|^p \, dt\right]^{1/p} = \left[\frac{1}{-h}\int_0^{-h} |t^k|^p \, dt\right]^{1/p}$$

$$= \left[\frac{1}{-h}\int_0^{-h} \xi^{kp} \, d\xi\right]^{1/p} = \left[\frac{(-h)^{kp}}{kp+1}\right]^{1/p} = O(h^k) \text{ as } h \to 0,$$

$$(1.12.5)$$

and so in any case

$$\frac{1}{k!}|f_{(k),p}(x)|\left[\frac{1}{h}\int_0^h |t^k|^p \, dt\right]^{1/p} = o(h^{k-1}) \text{ as } h \to 0. \qquad (1.12.6)$$

Now, by Minkowski's inequality

$$\left[\frac{1}{h}\int_0^h \left|f(x+t) - P(t) + \frac{t^k}{k!}f_{(k),p}(x)\right|^p \, dt\right]^{1/p}$$

$$\leq \left[\frac{1}{h}\int_0^h |f(x+t) - P(t)|^p \, dt\right]^{1/p} + \frac{1}{k!}|f_{(k),p}(x)|\left[\frac{1}{h}\int_0^h |t^k|^p \, dt\right]^{1/p}$$

and so by (1.12.1) and (1.12.6)

$$\left[\frac{1}{h}\int_0^h \left|f(x+t) - P(t) + \frac{t^k}{k!}f_{(k),p}(x)\right|^p \, dt\right]^{1/p} = o(h^{k-1}) \text{ as } h \to 0. \quad (1.12.7)$$

Since $P(t) - \frac{t^k}{k!}f_{(k),p}(x)$ is a polynomial of degree at most $k-1$, f has a $(k-1)$th L^p derivative at x. ☐

Thus, if $f_{(k),p}(x)$ exists, then all the previous derivatives $f_{(i),p}(x)$, $1 \leq i \leq k-1$ exist.

If a_0 is the constant term in $P(t)$, then writing $f_{(0),p}(x) = a_0$ we get, as in (1.12.7)

$$\left[\frac{1}{h}\int_0^h |f(x+t) - f_{(0),p}(x)|^p \, dt\right]^{1/p} = o(1) \text{ as } h \to 0.$$

Definition. A function f is said to be L^p-*continuous* at x if

$$\left[\frac{1}{h}\int_0^h |f(x+t) - f(x)|^p \, dt\right]^{1/p} = o(1) \text{ as } h \to 0.$$

If f is continuous at x, then f is L^p-continuous at x and, if f is L^p-continuous at x, then $f_{(0),p}(x)$ exists and $f_{(0),p}(x) = f(x)$. In particular, if $p = 1$, then f is L^p-continuous at x if and only if x is a Lebesgue point of f.

It can be shown that if $f \in L^p$, $p \geq 1$, then f is L^p-continuous almost everywhere [196; I, p. 65].

Lemma 1.12.1 *If* $0 < r < p < \infty$ *and* $f \in L^p$ *in* $[a, b]$, *then* $f \in L^r$ *in* $[a, b]$.

☐   If E_1 and E_2 are the sets of points x in $[a, b]$ where $|f(x)| \leq 1$ and where $|f(x)| > 1$, respectively, then

$$\int_a^b |f|^r = \int_{E_1} |f|^r + \int_{E_2} |f|^r \leq (b - a) + \int_a^b |f|^p < \infty,$$

and so the result follows.   ☐

Lemma 1.12.2 *If* $0 < r < p < \infty$ *and* $f \in L^p$ *in some neighbourhood of* x, *then for small* h

$$\left[\frac{1}{h} \int_0^h |f(x + t) - P(t)|^r \, dt \right]^{1/r} \leq \left[\frac{1}{h} \int_0^h |f(x + t) - P(t)|^p \, dt \right]^{1/p} \quad (1.12.8)$$

for any polynomial $P(t)$.

☐   By hypothesis, $f \in L^p$ and so $f(x + t) - P(t) \in L^p$. As a result, by Lemma 1.12.1, $f(x + t) - P(t) \in L^r$ in some neighbourhood of x. Hence, by Hölder's inequality, assuming $h > 0$,

$$\int_0^h |f(x + t) - P(t)|^r \, dt \leq h^{(p-r)/p} \left(\int_0^h |f(x + t) - P(t)|^p \, dt \right)^{r/p},$$

which is (1.12.8). If $h < 0$, taking the integrals from h to 0, we again get (1.12.8).   ☐

Theorem 1.12.3 *If* $0 < r < p < \infty$ *and if* f *has a* kth L^p-derivative at x, *then* f *has a* kth L^r-derivative at x *and they are equal*.

☐   If f has a kth L^p-derivative at x, then

$$\left(\frac{1}{h} \int_0^h \left| f(x + t) - \sum_{i=0}^k \frac{t^i}{i!} f_{(i),p}(x) \right|^p \, dt \right)^{1/p} = o(h^k) \text{ as } h \to 0,$$

and so by Lemma 1.12.2

$$\left(\frac{1}{h} \int_0^h \left| f(x + t) - \sum_{i=0}^k \frac{t^i}{i!} f_{(i),p}(x) \right|^r \, dt \right)^{1/r} = o(h^k) \text{ as } h \to 0.$$

Hence, $f_{(k),r}(x)$ exists and $f_{(i),r}(x) = f_{(i),p}(x)$ for $i = 0, 1, \ldots, k$.   ☐

1.12.2 L^p-Boundedness

If in (1.12.1) $o(h^k)$ is replaced by $O(h^k)$ as $h \to 0$, then f is said to be *L^p-bounded of order k at x.*

Theorem 1.12.4 *If f is L^p-bounded of order k at x, then $f_{(k-1),p}(x)$ exists.*

□   From the definition of L^p-boundedness, we have

$$\left(\frac{1}{h}\int_0^h |f(x+t) - Q(t)|^p \, dt\right)^{1/p} = O(h^k) \text{ as } h \to 0, \qquad (1.12.9)$$

where $Q(t)$ is a polynomial of degree at most k. Let $Q(t) = \sum_{i=0}^k b_i t^i$ and $\overline{Q}(t) = \sum_{i=0}^{k-1} b_i t^i$. We have seen in (1.12.4) and (1.12.5) that

$$\left(\frac{1}{h}\int_0^h |t|^k \, dt\right)^{1/p} = O(h^k) \text{ as } h \to 0. \qquad (1.12.10)$$

Applying Minkowski's inequality, we get from (1.12.9) and (1.12.10)

$$
\begin{aligned}
\frac{1}{h}\int_0^h |f(x+t) - \overline{Q}(t)|^p \, dt &= \left(\frac{1}{h}\int_0^h |f(x+t) - Q(t) + b_k t^k|^p \, dt\right)^{1/p} \\
&\leq \left(\frac{1}{h}\int_0^h |f(x+t) - Q(t)|^p \, dt\right)^{1/p} \\
&\quad + |b_k|\left(\frac{1}{h}\int_0^h |t|^k \, dt\right)^{1/p} \\
&= O(h^k) + O(h^k) = O(h^k) \text{ as } h \to 0.
\end{aligned}
$$
$$(1.12.11)$$

Since $O(h^k)$ implies $o(h^{k-1})$ as $h \to 0$, (1.12.11) shows that $f_{(k-1),p}(x)$ exists.
□

The argument also shows that $\overline{Q}(t) = \sum_{i=0}^{k-1} \frac{t^i}{i!} f_{(i),p}(x)$ and so, from (1.12.11), we get:

Corollary 1.12.5 *If f is L^p-bounded of order k at x, then $f_{(k-1),p}(x)$ exists and*

$$\left(\frac{1}{h}\int_0^h \left|f(x+t) - \sum_{i=0}^{k-1} \frac{t^i}{i!} f_{(i),p}(x)\right|^p \, dt\right)^{1/p} = O(h^k) \text{ as } h \to 0.$$

1.13   Symmetric L^p-Derivatives

1.13.1   Symmetric L^p-Derivatives and Symmetric L^p-Continuity

Assume that $f \in L^p$, $1 \leq p < \infty$, in some neighbourhood of x. If there is a polynomial $Q(t) = Q_x(t)$ of degree at most k for which

$$\left(\frac{1}{h}\int_0^h \left|\frac{f(x+t) + (-1)^k f(x-t)}{2} - Q(t)\right|^p dt\right)^{1/p} = o(h^k) \text{ as } h \to 0, \quad (1.13.1)$$

then f is said to have a *kth symmetric L^p-derivative* at x and, if $a_k/k!$ is the coefficient of t^k in $Q(t)$, then a_k is called the *kth symmetric L^p-derivative of f at x* and is denoted by $f_{(k),p}^{(s)}(x)$.

It can be shown as in the case of the unsymmetric L^p-derivative that if $f_{(k),p}^{(s)}(x)$ exists it is unique.

We show that if $f_{(k),p}^{(s)}(x)$ exists, then $Q(t)$ has only even or odd powers of t according as k is even or odd.

☐   Let $k = 2m$. Then for $h > 0$ we have from (1.13.1)

$$\left(\frac{1}{h}\int_0^h \left|\frac{f(x+t) + f(x-t)}{2} - Q(t)\right|^p dt\right)^{1/p} = o(h^{2m}) \text{ as } h \to 0. \quad (1.13.2)$$

This being true for all h, changing h to $-h$,

$$\left(\frac{1}{-h}\int_0^{-h} \left|\frac{f(x+t) + f(x-t)}{2} - Q(t)\right|^p dt\right)^{1/p} = o(h^{2m}) \text{ as } h \to 0.$$

Changing t to $-t$ this becomes

$$\left(\frac{1}{h}\int_0^h \left|\frac{f(x-t) + f(x+t)}{2} - Q(-t)\right|^p dt\right)^{1/p} = o(h^{2m}) \text{ as } h \to 0. \quad (1.13.3)$$

Therefore, applying Minkowski's inequality, we have from (1.13.2) and (1.13.3)

$$\left(\frac{1}{h}\int_0^h |Q(t) - Q(-t)|^p dt\right)^{1/p} = o(h^{2m}) \text{ as } h \to 0. \quad (1.13.4)$$

If we put $Q(t) = \sum_{i=0}^{2m} b_i t^i$, then $Q(t) - Q(-t) = 2\sum_{i=1}^m b_{2i-1} t^{2i-1}$ and so $\lim_{t\to 0} \frac{Q(t) - Q(-t)}{t} = 2b_1$. Therefore, if $b_1 \neq 0$, $\lim_{t\to 0}\left|\frac{Q(t) - Q(-t)}{t}\right| > 0$ and so there are $\epsilon > 0$ and $h > 0$ such that $\left|\frac{Q(t) - Q(-t)}{t}\right| > \epsilon$ for $0 < t < h$, which gives

$$\left(\frac{1}{h}\int_0^h |Q(t) - Q(-t)|^p dt\right)^{1/p} > \epsilon \frac{h}{(p+1)^{1/p}},$$

which contradicts (1.13.4). So, $b_1 = 0$.

Suppose that $b_1 = b_3 = \cdots = b_{2r-1} = 0$, where $r < m$. If $b_{2r+1} \neq 0$, then $\lim_{t \to 0} \left| \dfrac{Q(t) - Q(-t)}{t^{2r+1}} \right| > 0$ and so, as above, there is an $\epsilon > 0$ and an $h > 0$ such that

$$\left(\frac{1}{h} \int_0^h |Q(t) - Q(-t)|^p \, dt \right)^{1/p} > \epsilon \frac{h^{2r+1}}{\left(p(2r+1) + 1 \right)^{1/p}},$$

which contradicts (1.13.4) since $2r + 1 < 2m$. Hence, $b_{2r+1} = 0$ and so by induction $b_{2i-1} = 0$ for $i = 1, 2, \ldots, m$.

The case when k is odd is similar. In this case the integrand in (1.13.4) is $|Q(t) + Q(-t)|^p$ and the right-hand side of (1.13.4) is $o(h^{2m+1})$ from which we get ultimately that $Q(t) = -Q(-t)$. □

Also, if $f_{(k),p}^{(s)}(x)$ exists, then $f_{(k-2),p}^{(s)}(x)$ exists.

□   For, as in (1.12.4),

$$\frac{1}{k!} |f_{(k),p}^{(s)}(x)| \left(\frac{1}{h} \int_0^h |t^k|^p \, dt \right)^{1/p} = o(h^{k-1}) \text{ as } h \to 0, \tag{1.13.5}$$

and, applying Minkowski's inequality, we have from (1.13.1 and (1.13.5)

$$\left(\frac{1}{h} \int_0^h \left| \frac{f(x+t) + (-1)^k f(x-t)}{2} - Q(t) + \frac{t^k}{k!} f_{(k),p}^{(s)}(x) \right|^p \, dt \right)^{1/p}$$
$$= o(h^{k-2}) \text{ as } h \to 0.$$

Since $Q(t) - \frac{t^k}{k!} f_{(k),p}^{(s)}(x)$ is a polynomial in t of degree at most $k-2$, $f_{(k-2),p}^{(s)}(x)$ exists. □

Thus, if $f_{(k),p}^{(s)}(x)$ exists, then $f_{(k-2),p}^{(s)}(x)$, $f_{(k-4),p}^{(s)}(x)$, ..., exists. If k is even, the existence of $f_{(0),p}^{(s)}(x)$ means that $f_{(0),p}^{(s)}(x) = a_0$, where a_0 is the constant term in the polynomial $Q(t)$ in (1.13.1) and so

$$\left(\frac{1}{h} \int_0^h \left| \frac{f(x+t) + (-1)^k f(x-t)}{2} - f_{(0),p}^{(s)}(x) \right|^p \, dt \right)^{1/p} = o(1) \text{ as } h \to 0.$$

Definition. A function f is said to be *symmetric L^p-continuous at x of even order* if

$$\left(\frac{1}{h} \int_0^h \left| \frac{f(x+t) + f(x-t)}{2} - f(x) \right|^p \, dt \right)^{1/p} = o(1) \text{ as } h \to 0,$$

and f is said to be *symmetric L^p-continuous at x of odd order* if

$$\left(\frac{1}{h} \int_0^h \left| f(x+t) - f(x-t) \right|^p \, dt \right)^{1/p} = o(1) \text{ as } h \to 0.$$

Clearly, if f is symmetric continuous at x of even, (odd), order, then f is symmetric L^p-continuous at x of even, (odd), order.

1.13.2 L^p-Smoothness

Let $f \in L^p$, $1 \le p < \infty$, on some neighbourhood of x and let $f_{(k-2),p}^{(s)}(x)$ exist, where $k \ge 2$. Let

$$
P(t) = \begin{cases} \displaystyle\sum_{i=0}^{(k/2)-1} \frac{t^{2i}}{(2i)!} f_{(2i),p}^{(s)}(x), & \text{if } k \text{ is even}, \\[3mm] \displaystyle\sum_{i=0}^{(k-1)/2} \frac{t^{2i-1}}{(2i-1)!} f_{(2i-1),p}^{(s)}(x), & \text{if } k \text{ is odd}. \end{cases}
\tag{1.13.6}
$$

If

$$
\left[\frac{1}{h} \int_0^h \left| \frac{f(x+t) + (-1)^k f(x-t)}{2} - P(t) \right|^p dt \right]^{1/p} = o(h^{k-1}) \text{ as } h \to 0,
\tag{1.13.7}
$$

then f is said to be L^p-smooth of order k at x.

Theorem 1.13.1 *If $f_{(k),p}^{(s)}(x)$ exists, then f is L^p-smooth of order k at x.*

□   Let $f_{(k),p}^{(s)}(x)$ exist, then

$$
\left[\frac{1}{h} \int_0^h \left| \frac{f(x+t) + (-1)^k f(x-t)}{2} - P(t) - \frac{t^k}{k!} f_{(k),p}^{(s)}(x) \right|^p dt \right]^{1/p}
$$
$$
= o(h^k) \text{ as } h \to 0,
\tag{1.13.8}
$$

where $P(t)$ is as in (1.13.6). Applying Minkowski's inequality, we have from (1.13.8) and (1.12.10)

$$
\left[\frac{1}{h} \int_0^h \left| \frac{f(x+t) + (-1)^k f(x-t)}{2} - P(t) \right|^p dt \right]^{1/p}
$$
$$
= \left[\frac{1}{h} \int_0^h \left| \frac{f(x+t) + (-1)^k f(x-t)}{2} - P(t) - \frac{t^k}{k!} f_{(k),p}^{(s)}(x) + \frac{t^k}{k!} f_{(k),p}^{(s)}(x) \right|^p dt \right]^{1/p}
$$
$$
\le \left[\frac{1}{h} \int_0^h \left| \frac{f(x+t) + (-1)^k f(x-t)}{2} - P(t) - \frac{t^k}{k!} f_{(k),p}^{(s)}(x) \right|^p dt \right]^{1/p}
$$
$$
+ \frac{1}{h} \left[\int_0^h \left| \frac{t^k}{k!} f_{(k),p}^{(s)}(x) \right|^p dt \right]^{1/p}
$$
$$
= o(h^k) + O(h^k) \text{ as } h \to 0.
\tag{1.13.9}
$$

From (1.13.9), we conclude that (1.13.7) is satisfied and so the result follows.
□

Corollary 1.13.2 *If f is L^p-smooth of order k at x, then f is L^p-smooth of order $k - 2$ at x.*

□   If f is L^p-smooth of order k at x, then by definition $f_{(k-2),p}^{(s)}(x)$ exists and so the result follows from Theorem 1.13.1.   □

1.13.3 Symmetric L^p-Boundedness

If in (1.13.1), $o(h^k)$ is replaced by $O(h^k)$ as $h \to 0$, then f is said to be *symmetric L^p-bounded at x of order k.*

Theorem 1.13.3 *If f is symmetric L^p-bounded at x of order k, then $f^{(s)}_{(k-2),p}(x)$ exists.*

☐   Since f is symmetric L^p-bounded at x of order k,

$$\left[\frac{1}{h}\int_0^h \left|\frac{f(x+t)+(-1)^k f(x-t)}{2} - Q(t)\right|^p dt\right]^{1/p} = O(h^k) \text{ as } h \to 0,$$
(1.13.10)

where $Q(t)$ is a polynomial of degree at most k. Let $Q(t) = \sum_{i=0}^k b_i t^i$ and $\overline{Q}(t) = \sum_{i=0}^{k-2} b_i t^i$. As in (1.12.10),

$$\left[\frac{1}{h}\int_0^h |t^k|^p dt\right]^{1/p} = O(h^k), \quad \left[\frac{1}{h}\int_0^h |t^{k-1}|^p dt\right]^{1/p} = O(h^{k-1}) \text{ as } h \to 0.$$
(1.13.11)

Applying Minkowski's inequality, we have from (1.13.10) and (1.13.11)

$$\left[\frac{1}{h}\int_0^h \left|\frac{f(x+t)+(-1)^{k-2} f(x-t)}{2} - \overline{Q}(t)\right|^p dt\right]^{1/p}$$

$$\leq \left[\frac{1}{h}\int_0^h \left|\frac{f(x+t)+(-1)^{k-2} f(x-t)}{2} - Q(t)\right|^p dt\right]^{1/p}$$

$$+|b_{k-1}|\left[\frac{1}{h}\int_0^h |t^{k-1}|^p dt\right]^{1/p} + |b_k|\left[\frac{1}{h}\int_0^h |t^k|^p dt\right]^{1/p}$$

$$= O(h^k) + O(h^{k-1}) + O(h^k) \text{ as } h \to 0.$$
(1.13.12)

Since the right-hand side of (1.13.12) implies $o(h^{k-2})$ as $h \to 0$, it follows that $f^{(s)}_{(k-2),p}(x)$ exists.   ☐

1.14   Abel Derivatives

1.14.1   Abel Summability

The concept of the Abel derivative of a function f at a point x is closely related to the concept of Abel summability at x of the differentiated series of the Fourier series of f. So, we first discuss Abel summability and deduce some results needed in the sequel.

Let $\sum_{n=0}^\infty u_n$ be a series of constant terms such that $\sum_{n=0}^\infty u_n r^n$ converges for each r, $0 < r < 1$, and let $\phi(r) = \sum_{n=0}^\infty u_n r^n$. Then $\limsup_{r\to 1} \phi(r)$

and $\liminf_{r \to 1} \phi(r)$ are respectively called the *upper* and *lower Abel sums* of $\sum_{n=0}^{\infty} u_n$. If the upper and lower Abel sums of $\sum_{n=0}^{\infty} u_n$ are equal and finite, then the series $\sum_{n=0}^{\infty} u_n$ is called *Abel summable* and the common value is called the *Abel sum* of $\sum_{n=0}^{\infty} u_n$. It is clear that if $\sum_{n=0}^{\infty} u_n$ is convergent, then it is Abel summable and the Abel sum is equal to the sum of the series $\sum_{n=0}^{\infty} u_n$.

Lemma 1.14.1 *If the upper and lower Abel sums of $\sum_{n=0}^{\infty} u_n$ are finite, then the series $\sum_{n=1}^{\infty} u_n/n^m$ is Abel summable for all integers $m \geq 1$.*

☐   Let $m = 1$. Since $\sum_{n=0}^{\infty} u_n$ has finite upper and lower Abel sums, the series $\sum_{n=0}^{\infty} u_n r^n$ converges for each r, $0 < r < 1$, and $\sum_{n=0}^{\infty} u_n r^n$ remains bounded as $r \to 1-$. So, the power series $\sum_{n=0}^{\infty} u_n r^n$ has a radius of convergence of at least 1. Let

$$\phi(r) = \sum_{n=0}^{\infty} u_n r^n, \quad \Phi(r) = \sum_{n=1}^{\infty} \frac{u_n}{n} r^n.$$

Then

$$\int_0^r \frac{\phi(t) - u_0}{t}\, dt = \int_0^r \sum_{n=1}^{\infty} u_n t^{n-1}\, dt = \sum_{n=1}^{\infty} \frac{u_n}{n} r^n = \Phi(r), \ 0 < r < 1.$$

Since $\phi(r)$ is bounded as $r \to 1-$, $\dfrac{\phi(t) - u_0}{t}$ is bounded as $t \to 1-$. Hence,

$$|\Phi(r) - \Phi(r')| = \left| \int_r^{r'} \frac{\phi(t) - u_0}{t}\, dt \right| \to 0 \text{ as } r, r' \to 1-.$$

So, $\lim_{r \to 1-} \Phi(r)$ exists finitely, which shows that $\sum_{n=1}^{\infty} u_n/n$ is Abel summable.

If $m \geq 1$, then applying this result for $m = 1$ repeatedly for $m - 1$ times completes the proof.   ☐

1.14.2   Abel Derivatives

Let f be a 2π-periodic Lebesgue integrable function and let

$$a_n = \frac{1}{\pi} \int_{-\pi}^{\pi} f(t) \cos nt\, dt, \quad b_n = \frac{1}{\pi} \int_{-\pi}^{\pi} f(t) \sin nt\, dt \qquad (1.14.1)$$

be the Fourier coefficients of f. By the Riemann–Lebesgue theorem $a_n \to 0$ and $b_n \to 0$ as $n \to \infty$ and, so, there is an $M > 0$ such that $|a_n| \leq M$, $|b_n| \leq M$ for all n. Hence,

$$\left| (a_n \cos nx + b_n \sin nx) r^n \right| \leq 2Mr^n \text{ for all } n \text{ and } x, \ 0 < r < 1. \qquad (1.14.2)$$

Thus, the series $\frac{1}{2}a_0 + \sum_{n=1}^{\infty}(a_n \cos nx + b_n \sin nx)r^n$ converges uniformly and absolutely for fixed r, $0 < r < 1$, and let $f(r,x)$ be its sum. Then, using (1.14.1),

$$f(r,x) = \frac{1}{2}a_0 + \sum_{n=1}^{\infty}(a_n \cos nx + b_n \sin nx)r^n$$

$$= \frac{1}{2\pi}\int_{-\pi}^{\pi} f$$

$$+ \sum_{n=1}^{\infty} r^n \left(\left(\frac{1}{\pi}\int_{-\pi}^{\pi} f(t) \cos nt \, dt\right) \cos nx + \left(\frac{1}{\pi}\int_{-\pi}^{\pi} f(t) \sin nt \, dt\right) \sin nx \right)$$

$$= \frac{1}{\pi}\int_{-\pi}^{\pi} f(t) \left(\frac{1}{2} + \sum_{n=1}^{\infty} r^n \cos n(x-t) \right) dt$$

$$= \frac{1}{\pi}\int_{-\pi}^{\pi} f(x+t) \left(\frac{1}{2} + \sum_{n=1}^{\infty} r^n \cos nt \right) dt$$

$$= \frac{1}{\pi}\int_{-\pi}^{\pi} f(x+t)P(r,t) \, dt, \qquad (1.14.3)$$

where

$$P(r,t) = \frac{1}{2} + \sum_{n=1}^{\infty} r^n \cos nt. \qquad (1.14.4)$$

Clearly the interchange of the order of summation and integration and the change of variable are justified. Now consider

$$\frac{1}{2} + z + z^2 + z^3 \cdots = \frac{1}{2}\frac{1+z}{1-z}, \quad z = re^{it}, \ 0 < r < 1.$$

Equating its real part, we have

$$\frac{1}{2} + \sum_{n=1}^{\infty} r^n \cos nt = \frac{1}{2}\left[\frac{1-r^2}{1-2r\cos t + r^2}\right]$$

and so

$$P(r,t) = \frac{1}{2}\left[\frac{1-r^2}{1-2r\cos t + r^2}\right] = \frac{1}{2}\left[\frac{1-r^2}{(1-r)^2 + 4r\sin^2 t/2}\right]. \qquad (1.14.5)$$

So, we have

$$P(r,t) \geq 0; \ P(r,-t) = P(r,t); \qquad (1.14.6)$$

$$\frac{1}{\pi}\int_{-\pi}^{\pi} P(r,t) \, dt = 1; \qquad (1.14.7)$$

$$P(r,t) < \frac{1}{1-r}; \qquad (1.14.8)$$

$$P(r,t) \leq C\frac{1-r}{t^2} \ \text{if} \ \frac{1}{2} \leq r < 1 \ \text{and} \ 0 < |t| \leq \pi, C \ \text{being a constant.} (1.14.9)$$

☐   In fact, (1.14.6) follows from (1.14.5) and (1.14.7) follows from (1.14.4) using term-by-term integration. From (1.14.5),

$$P(r,t) \leq \frac{1}{2}\frac{1-r^2}{(1-r)^2} = \frac{1}{2}\frac{1-r}{1+r} < \frac{1}{1-r},$$

proving (1.14.8). Finally, $(t\cos t - \sin t)' < 0$ if $0 < t < \pi$ and, so, $t\cos t - \sin t$ is decreasing on that interval. Thus, if $0 < t < \pi$, $t\cos t - \sin t < 0$. Hence, $(\sin t/t)' = (t\cos t - \sin t)/t^2 < 0$ if $0 < t < \pi$ showing that $\sin t/t$ is decreasing if $0 < t < \pi$. Hence, $\sin t \geq 2t/\pi$ if $0 \leq t \leq \pi/2$ and, so, from (1.14.5) if $\frac{1}{2} \leq r < 1$ and $|t| \leq \pi$,

$$P(r,t) \leq \frac{1}{2}\frac{1-r^2}{4r\sin^2 t/2} \leq \frac{1-r}{2\sin^2 t/2} \leq \frac{\pi^2}{2}\frac{1-r}{t^2},$$

proving (1.14.9).   ☐

Lemma 1.14.2 *If f is a 2π-periodic Lebesgue integrable function and if at a point x, $\lim_{t\to 0+} [f(x+t) + f(x-t) - 2f(x)] = 0$, then the Fourier series of f is Abel summable at x to $f(x)$.*

☐   Let $\epsilon > 0$ be arbitrary. Then there is a $\delta > 0$, $0 < \delta < \pi$, such that $|f(x+t) + f(x-t) - 2f(x)| < \epsilon$ whenever $0 < t < \delta$. So, from (1.14.6) and (1.14.7)

$$\left|\frac{1}{\pi}\int_0^\delta [f(x+t) + f(x-t) - 2f(x)] P(r,t)\,dt\right| \leq \frac{\epsilon}{\pi}\int_0^\delta P(r,t)\,dt$$

$$\leq \frac{\epsilon}{\pi}\int_{-\pi}^\pi P(r,t)\,dt = \epsilon.$$
$$(1.14.10)$$

Also, from (1.14.9), taking $\frac{1}{2} \leq r < 1$, we have

$$\left|\frac{1}{\pi}\int_\delta^\pi [f(x+t) + f(x-t) - 2f(x)] P(r,t)\,dt\right|$$

$$\leq C\frac{1-r}{\pi}\int_\delta^\pi \frac{|f(x+t) + f(x-t) - 2f(x)|}{t^2}\,dt$$

$$\leq C\frac{1-r}{\delta^2\pi}\int_\delta^\pi |f(x+t) + f(x-t) - 2f(x)|\,dt,$$

and, thus,

$$\lim_{r\to 1-}\left|\frac{1}{\pi}\int_\delta^\pi [f(x+t) + f(x-t) - 2f(x)] P(r,t)\,dt\right| = 0.   (1.14.11)$$

From (1.14.3), (1.14.7) and (1.14.6)

$$f(r,x) - f(x) = \frac{1}{\pi} \int_{-\pi}^{\pi} [f(x+t) - f(x)] P(r,t) \, dt$$

$$= \frac{1}{\pi} \left(\int_{-\pi}^{0} + \int_{0}^{\pi} \right) [f(x+t) - f(x)] P(r,t) \, dt$$

$$= \frac{1}{\pi} \int_{0}^{\pi} [f(x+t) + f(x-t) - 2f(x)] P(r,t) \, dt$$

$$= \frac{1}{\pi} \left(\int_{0}^{\delta} + \int_{\delta}^{\pi} \right) [f(x+t) + f(x-t) - 2f(x)] P(r,t) \, dt$$

$$(1.14.12)$$

and so from (1.14.10), (1.14.11) and (1.14.12), $\limsup_{r \to 1-} |f(r,x) - f(x)| \le \epsilon$. Since ϵ was arbitrary, it follows that $\lim_{r \to 1-} f(r,x) = f(x)$, completing the proof. □

Let k be a positive integer. If k is even, then from (1.14.2)

$$\left| (-1)^{k/2} (a_n n^k \cos nx + b_n n^k \sin nx) r^n \right| \le 2Mn^k r^n \text{ for all } x \text{ and } n, \ 0 < r < 1.$$

Since $\lim_{n \to \infty} (n^k r^n)^{1/n} = r < 1$, the series $\sum_{n=1}^{\infty} n^k r^n$ is convergent and so the series $\sum_{n=1}^{\infty} (-1)^{k/2} (a_n n^k \cos nx + b_n n^k \sin nx) r^n$ is uniformly and absolutely convergent for fixed r, $0 < r < 1$. Hence, the series

$$f(r,x) = \frac{1}{2} a_0 + \sum_{n=1}^{\infty} (a_n \cos nx + b_n \sin nx) r^n \qquad (1.14.13)$$

can be differentiated term-by-term k times with respect to x.

If k is odd, then as in (1.14.2)

$$\left| (a_n \sin nx - b_n \cos nx) r^n \right| \le 2Mr^n \text{for all } n \text{ and } x, \ 0 < r < 1,$$

and so

$$\left| (-1)^{(k+1)/2} (a_n n^k \sin nx - b_n n^k \cos nx) r^n \right| \le 2Mn^k r^n$$
$$\text{for all } x \text{ and } n, \ 0 < r < 1,$$

and, as above, we can differentiate the series (1.14.13) term-by-term k times with respect to x.

So differentiating (1.14.13) k times with respect to x, we have

$$\frac{\partial^k f}{\partial x^k}(r,x) = \qquad\qquad\qquad (1.14.14)$$

$$\begin{cases} (-1)^{k/2} \sum_{n=1}^{\infty} (a_n n^k \cos nx + b_n n^k \sin nx) r^n, & \text{if } k \text{ is even,} \\ (-1)^{(k+1)/2} \sum_{n=1}^{\infty} (a_n n^k \sin nx - b_n n^k \cos nx) r^n, & \text{if } k \text{ is odd.} \end{cases}$$

The *upper and lower Abel derivates of f at x of order k* are defined by

$$\overline{AD}_k f(x) \;=\; \limsup_{r \to 1-} \frac{\partial^k f}{\partial x^k}(r, x),$$

$$\underline{AD}_k f(x) \;=\; \liminf_{r \to 1-} \frac{\partial^k f}{\partial x^k}(r, x),$$

respectively. If $\overline{AD}_k f(x) = \underline{AD}_k f(x)$, the common value is called the *Abel derivative of f at x of order k* and is denoted by $AD_k f(x)$.

Theorem 1.14.3 *If $\overline{AD}_k f(x)$ and $\underline{AD}_k f(x)$ are finite and $k \geq 2$, then $AD_{k-2} f(x)$ exists finitely.*

☐   The proof follows from Lemma 1.14.1.   ☐

While the finiteness of $\overline{AD}_k f(x)$ and $\underline{AD}_k f(x)$ implies the existence of $AD_{k-2} f(x)$ finitely, nothing can be said about the existence of $AD_{k-1} f(x)$ since Abel summability at x does not ensure the Abel summability of the once integrated series at the point x.

Example. The series $\sum_{n=1}^{\infty} \sin nx$ is Abel summable at $x = 0$, but the series $\sum_{n=1}^{\infty} (\cos nx)/n$ is not Abel summable at $x = 0$.

Because of this the Abel derivative is considered as a symmetric derivative.

From Theorem 1.14.3, it follows that if $AD_k f(x)$ exists finitely, then $AD_{k-2} f(x)$ exists finitely and so $AD_0 f(x)$ or $AD_1 f(x)$ exists finitely according as k is even or odd, where $AD_0 f(x) = \lim_{r \to 1-} f(r, x)$. Thus, the Abel derivative at x of order 0 is just the Abel sum at x of the Fourier series of f. From Lemma 1.14.2, if f is continuous at x, then $AD_0 f(x) = f(x)$.

1.14.3   Abel Continuity

Let f be a 2π-periodic Lebesgue integrable function and let $f(r, x)$ be as in (1.14.3). Then $\limsup_{r \to 1-} f(r, x)$ and $\liminf_{r \to 1-} f(r, x)$ are respectively called the *upper and lower Abel limits of f at x*. If $\limsup_{r \to 1-} f(r, x) = \liminf_{r \to 1-} f(r, x)$, then this common value is called the *Abel limit if f at x*. If the Abel limit exists finitely with value $f(x)$, then f is said to be *Abel continuous at x*. The upper and lower Abel sums of the Fourier series of f at x are the upper and lower Abel limits of f at x, respectively. Also the Fourier series of f is Abel summable to S at x if and only if the Abel limit of f at x exists and is equal to S, and the Fourier series of f is Abel summable to $f(x)$ at x if and only if f is Abel continuous at x. In the following, we shall see that if f is continuous at x, then f is Abel continuous at x. The converse is obviously not true.

Theorem 1.14.4 *Let f be 2π-periodic and Lebesgue integrable. Then for any point x*

$$\liminf_{t\to 0} \frac{f(x+t)+f(x-t)}{2} \leq \liminf_{r\to 1-} f(r,x) \leq \limsup_{r\to 1-} f(r,x)$$

$$\leq \limsup_{t\to 0} \frac{f(x+t)+f(x-t)}{2}.$$

☐   We prove the right-hand side inequality, the proof for left-hand side inequality being similar.

Let $\epsilon > 0$ be arbitrary and $L = \limsup_{t\to 0} \dfrac{f(x+t)+f(x-t)}{2}$. Then there exists δ, $0 < \delta < \pi$, such that $\dfrac{f(x+t)+f(x-t)}{2} < L + \epsilon$ for $0 < t < \delta$. So, from (1.14.6) and (1.14.7),

$$\frac{1}{\pi}\int_0^\delta [f(x+t)+f(x-t)]P(r,t)\,dt \leq \frac{(2L+2\epsilon)}{\pi}\int_0^\delta P(r,t)\,dt$$

$$\leq \frac{2(L+\epsilon)}{\pi}\int_0^\pi P(r,t)\,dt = \frac{(L+\epsilon)}{\pi}\int_{-\pi}^\pi P(r,t)\,dt = L+\epsilon. \quad (1.14.15)$$

From (1.14.9), taking $1/2 < r < 1$,

$$\frac{1}{\pi}\int_\delta^\pi [f(x+t)+f(x-t)]P(r,t)\,dt \leq \frac{1}{\pi}\int_\delta^\pi |f(x+t)+f(x-t)|P(r,t)\,dt$$

$$\leq \frac{1}{\pi}\int_\delta^\pi |f(x+t)+f(x-t)|C\frac{1-r}{t^2}\,dt$$

$$\leq \frac{C(1-r)}{\pi\delta^2}\int_\delta^\pi |f(x+t)+f(x-t)|\,dt$$

$$\leq \frac{C(1-r)}{\pi\delta^2}\int_0^\pi |f(x+t)+f(x-t)|\,dt.$$

$$(1.14.16)$$

From (1.14.15) and (1.14.16),

$$\frac{1}{\pi}\int_0^\pi [f(x+t)+f(x-t)]P(r,t)\,dt \leq L+\epsilon+\frac{C(1-r)}{\pi\delta^2}\int_0^\pi |f(x+t)+f(x-t)|\,dt.$$

Letting $r \to 1-$, we have

$$\limsup_{r\to 1-} \frac{1}{\pi}\int_0^\pi [f(x+t)+f(x-t)]P(r,t)\,dt \leq L+\epsilon. \quad (1.14.17)$$

Also, from (1.14.3) and (1.14.6), we have

$$f(r,x) = \frac{1}{\pi}\int_{-\pi}^{\pi} f(x+t)P(r,t)\,dt = \frac{1}{\pi}\int_{-\pi}^{\pi} f(x-t)P(r,t)\,dt$$

$$= \frac{1}{2\pi}\int_{-\pi}^{\pi}\big[f(x+t)+f(x-t)\big]P(r,t)\,dt$$

$$= \frac{1}{\pi}\int_{0}^{\pi}\big[f(x+t)+f(x-t)\big]P(r,t)\,dt. \qquad (1.14.18)$$

Letting $r \to 1-$ in (1.14.18), we have from (1.14.17) that $\limsup_{r\to 1-} f(r,x) \le L+\epsilon$. Since ϵ is arbitrary, the proof is complete. $\qquad\square$

Corollary 1.14.5 *If f is continuous at x then f is Abel continuous at x.*

1.14.4   Abel Smoothness

Let f be a 2π-periodic Lebesgue integrable function and let k be a fixed positive integer. Define

$$\overline{AS}_k f(x) = \limsup_{r\to 1-}(1-r)\frac{\partial^k f}{\partial x^k}(r,x)$$

$$\underline{AS}_k f(x) = \liminf_{r\to 1-}(1-r)\frac{\partial^k f}{\partial x^k}(r,x)$$

where $\dfrac{\partial^k f}{\partial x^k}(r,x)$ is as in (1.14.14). Then $\overline{AS}_k f(x)$ and $\underline{AS}_k f(x)$ are called the upper and lower indices of Abel smoothness of f of order k at x, respectively. If $\overline{AS}_k f(x) = \underline{AS}_k f(x)$, the common value, denoted by $AS_k f(x)$, is called the index of Abel smoothness of f of order k at x. If $\overline{AS}_k f(x) = \underline{AS}_k f(x) = 0$, then f is said to be *Abel smooth of order k at x*.

Theorem 1.14.6 *If $\overline{AD}_k f(x)$ and $\underline{AD}_k f(x)$ are finite, then f is Abel smooth of order k at x.*

$\square$ Since $\overline{AD}_k f(x)$ and $\underline{AD}_k f(x)$ are finite, $\dfrac{\partial^k f}{\partial x^k}(r,x)$ remains bounded as $r \to 1-$ and, so, $\overline{AS}_k f(x) = \underline{AS}_k f(x) = 0$, completing the proof. $\qquad\square$

1.15   Laplace Derivatives

1.15.1   Laplace Derivatives and Laplace Continuity

Let f be special Denjoy integrable in some neighbourhood $N(x)$ of x and let r be a fixed positive integer. If there exists a polynomial $Q(t) = Q_x(t)$ of degree at most r such that

$$s^{r+1} \int_0^\delta e^{-st} \left[f(x+t) - Q(t) \right] dt = o(1) \text{ as } s \to \infty \qquad (1.15.1)$$

for some $\delta > 0$ such that $x + \delta \in N(x)$, then f is said to have a *right Laplace derivative at x of order r* and if $a_r/r!$ is the coefficient of t^r in $Q(t)$, then a_r is called the *right Laplace derivative of f at x of order r* and is denoted by $LD_r^+ f(x)$. Note that if $\delta_1 > 0$, $\delta_2 > 0$, $\delta_1 \neq \delta_2$, $x + \delta_1, x + \delta_2 \in N(x)$, then integrating by parts,

$$s^{r+1} \int_{\delta_1}^{\delta_2} e^{-st} \left[f(x+t) - Q(t) \right] dt = s^{r+1} \left[e^{-s\delta_2} \int_{\delta_1}^{\delta_2} \left[f(x+t) - Q(t) \right] dt \right.$$

$$\left. + s \int_{\delta_1}^{\delta_2} e^{-st} \int_{\delta_1}^t \left[f(x+\xi) - Q(\xi) \right] d\xi \, dt \right]$$

$$= o(1) \text{ as } s \to \infty,$$

and, so, if (1.15.1) is true for $\delta = \delta_1$, then it is also true for $\delta = \delta_2$ and, therefore, (1.15.1) does not depend on δ.

If (1.15.1) is replaced by

$$s^{r+1} \int_{-\delta}^0 e^{st} \left[f(x+t) - Q(t) \right] dt = o(1) \text{ as } s \to \infty, \qquad (1.15.1')$$

or equivalently

$$s^{r+1} \int_0^\delta e^{-st} \left[f(x-t) - Q(-t) \right] dt = o(1) \text{ as } s \to \infty,$$

then f is said to have a *left Laplace derivative at x of order r* and if $a_r/r!$ is the coefficient of t^r in $Q(t)$, then a_r is called the *left Laplace derivative of f at x of order r* and is denoted by $LD_r^- f(x)$.

We shall consider $LD_r^+ f(x)$; the case for $LD_r^- f(x)$ is similar. We need the following lemma.

Lemma 1.15.1 *If $\delta > 0$ and p, q are positive integers, then*

$$s^q \int_0^\delta e^{-st} t^p \, dt = p! s^{q-p-1} + o(1) \text{ as } s \to \infty.$$

◻   Putting $\tau = st$,

$$s^q \int_0^\delta e^{-st} t^p \, dt = s^{q-p-1} \int_0^{s\delta} e^{-\tau} \tau^p \, d\tau$$

$$= s^{q-p-1} \int_0^\infty e^{-\tau} \tau^p \, d\tau - s^{q-p-1} \int_{s\delta}^\infty e^{-\tau} \tau^p \, d\tau$$

$$= p! s^{q-p-1} - s^{q-p-1} \int_{s\delta}^\infty e^{-\tau} \tau^p \, d\tau.$$

Let k be a positive integer such that $q - p + k \geq 2$. Then, since $\dfrac{\tau^{q+k}}{(q+k)!} < e^{\tau}$, we have

$$\int_{s\delta}^{\infty} e^{-\tau}\tau^p \, d\tau < \int_{s\delta}^{\infty} \frac{(q+k)!}{\tau^{q-p+k}} \, d\tau$$

$$= \frac{(q+k)!}{(q-p+k-1)(s\delta)^{q-p+k-1}} = o(1) \text{ as } s \to \infty,$$

completing the proof. □

Theorem 1.15.2 *The Laplace derivative $LD_r^+ f(x)$, if it exists, is unique.*

□   Let $Q_1(t) = \sum_{i=0}^{r} c_i t^i, Q_2(t) = \sum_{i=0}^{r} d_i t^i$ be two polynomials for which (1.15.1) is true. Then, by the above lemma

$$s^{r+1} \int_0^{\delta} e^{-st}\left[Q_1(t) - Q_2(t)\right] dt = \sum_{i=0}^{r}(c_i - d_i)s^{r+1}\int_0^{\delta} e^{-st}t^i \, dt$$

$$= \sum_{i=0}^{r}(c_i - d_i)i!s^{r-i} + o(1) \text{ as } s \to \infty.$$

$$(1.15.2)$$

Also, by (1.15.1),

$$s^{r+1} \int_0^{\delta} e^{-st}\left[Q_1(t) - Q_2(t)\right] dt = s^{r+1} \int_0^{\delta} e^{-st}\left[f(x+t) - Q_2(t)\right] dt$$

$$- s^{r+1} \int_0^{\delta} e^{-st}\left[f(x+t) - Q_1(t)\right] dt$$

$$= o(1) \text{ as } s \to \infty. \quad (1.15.3)$$

From (1.15.2) and (1.15.3), $\sum_{i=0}^{r}(c_i - d_i)i!s^{r-i} = o(1)$ as $s \to \infty$. Dividing by s^r and letting $s \to \infty$, we have $c_0 = d_0$. Hence, $\sum_{i=1}^{r}(c_i - d_i)i!s^{r-i} = o(1)$ as $s \to \infty$. Dividing by s^{r-1} and letting $s \to \infty$, we have $c_1 = d_1$.

Continuing, we have $c_i = d_i$, $i = 0, 1, \ldots, r-1$, and ultimately $(c_r - d_r)r! = o(1)$ as $s \to \infty$, from which we conclude that $c_r = d_r$, completing the proof. □

If $LD_r^+ f(x)$ exists, then $LD_i^+ f(x)$ exists for $i = 1, 2, \ldots, r-1$.

□   For suppose $LD_r^+ f(x)$ exists and let $a^j/j!$ be the coefficient of t^j in $Q(t)$, $0 \leq j \leq r$. If $\overline{Q}(t) = Q(t) - \sum_{j=i+1}^{r}(a^j/j!)t^j$, then, from (1.15.1) and from Lemma 1.15.1,

$$s^{i+1} \int_0^\delta e^{-st} [f(x+t) - \overline{Q}(t)] \, dt = s^{i+1} \Big[\int_0^\delta e^{-st} [f(x+t) - Q(t)] \, dt$$

$$+ \int_0^\delta e^{-st} \Big[\sum_{j=i+1}^r (a_j/j!) t^j \Big] \, dt \Big]$$

$$= \frac{1}{s^{r-i}} s^{r+1} \int_0^\delta e^{-st} [f(x+t) - Q(t)] \, dt$$

$$+ \sum_{j=i+1}^r \Big[\frac{a_j}{j!} s^{i+1} \int_0^\delta e^{-st} t^j \, dt \Big]$$

$$= \frac{1}{s^{r-i}} o(1) + \sum_{j=i+1}^r \Big[\frac{a_j}{j!} j! s^{i+1-j-1} + o(1) \Big]$$

$$\text{as } s \to \infty$$

$$= o(1) \text{ as } s \to \infty. \tag{1.15.4}$$

So, $\lim_{s\to\infty} s^{i+1} \int_0^\delta e^{-st} [f(x+t) - \overline{Q}(t)] \, dt = 0$ and, hence, (1.15.1) is satisfied if $Q(t)$ and r are replaced by $\overline{Q}(t)$ and i, respectively. Since $\overline{Q}(t)$ is a polynomial of degree at most i, $LD_i^+ f(x)$ exists and $LD_i^+ f(x) = a_i$. □

Since (1.15.4) is also true for $i = 0$, we have $s \int_0^\delta e^{-st} [f(x+t) - a_0] \, dt = o(1)$ as $s \to \infty$. Since $s \int_0^\delta e^{-st} \, dt \to 1$ as $s \to \infty$, writing $LD_0^+ f(x) = a_0$, we have

$$LD_0^+ f(x) = \lim_{s\to\infty} s \int_0^\delta e^{-st} f(x+t) \, dt, \tag{1.15.5}$$

which, as we now prove, is $f(x)$ if f is right-continuous at x.
□   Let $\epsilon > 0$ be arbitrary. Then there is a $\delta > 0$ such that $|f(x+t) - f(x)| < \epsilon$ for $0 < t < \delta$ and, therefore,

$$\Big| s \int_0^\delta e^{-st} [f(x+t) - f(x)] \, dt \Big| \le \epsilon s \int_0^\delta e^{-st} \, dt \to \epsilon \text{ as } s \to \infty. \tag{1.15.6}$$

So, from (1.15.5) and (1.15.6),

$$\big| LD_0^+ f(x) - f(x) \big|$$

$$= \Big| \lim_{s\to\infty} \Big[s \int_0^\delta e^{-st} [f(x+t) - f(x)] \, dt + s \int_0^\delta e^{-st} f(x) \, dt \Big] - f(x) \Big|$$

$$\le \epsilon.$$

Since ϵ is arbitrary the assertion follows. □

To define the *right upper and lower Laplace derivates of f at x of order k*, we suppose that $LD_{k-1}^+ f(x)$ exists finitely and let $Q(t) = \sum_{i=0}^{k-1} (t^i/i!) LD_i^+ f(x)$. Define

$$\overline{LD}_k^+ f(x) \;=\; \limsup_{s \to \infty} s^{k+1} \int_0^\delta e^{-st} \big[f(x+t) - Q(t) \big] \, dt; \qquad (1.15.7)$$

$$\underline{LD}_k^+ f(x) \;=\; \liminf_{s \to \infty} s^{k+1} \int_0^\delta e^{-st} \big[f(x+t) - Q(t) \big] \, dt. \qquad (1.15.8)$$

If $\overline{LD}_k^+ f(x) = \underline{LD}_k^+ f(x)$, the common value, possibly infinite, is the derivative $LD_k^+ f(x)$. This definition agrees with the former definition, as we now show.

□ Let $LD_k^+ f(x)$ exist in this sense and be finite. Then since, by Lemma 1.15.1, $\lim_{s \to \infty} s^{k+1} \int_0^\delta e^{-st}(t^k/k!) \, dt = 1$ and since, from (1.15.7) and (1.15.8),

$$\lim_{s \to \infty} s^{k+1} \int_0^\delta e^{-st} \Big[f(x+t) - \sum_{i=0}^{k-1} (t^i/i!) LD_i^+ f(x) \Big] \, dt = LD_k^+ f(x),$$

we have

$$\lim_{s \to \infty} s^{k+1} \int_0^\delta e^{-st} \Big[f(x+t) - \sum_{i=0}^{k} (t^i/i!) LD_i^+ f(x) \Big] \, dt = 0,$$

showing that $LD_k^+ f(x)$ exists in the former sense with the same value. The converse is clear. □

Definition A special Denjoy integrable function f is said to be *Laplace right-continuous at x* if

$$\lim_{s \to \infty} s \int_0^\delta e^{-st} f(x+t) \, dt = f(x)$$

for an arbitrary $\delta > 0$, and f is said to be *Laplace left-continuous at x* if

$$\lim_{s \to \infty} s \int_0^\delta e^{-st} f(x-t) \, dt = f(x)$$

for an arbitrary $\delta > 0$. If f is both Laplace left- and right-continuous at x, then f is said to be *Laplace continuous at x*.

We now show that if f is right-continuous, left-continuous, at x, then it is Laplace right-continuous, Laplace left-continuous, at x.

□ Suppose that f is right-continuous at x and let $\epsilon > 0$ be arbitrary. Then there is a $\delta > 0$ such that $\big| f(x+t) - f(x) \big| < \epsilon$ for $0 < t < \delta$. So,

$$\left| s \int_0^\delta e^{-st} \big[f(x+t) - f(x) \big] \, dt \right| \leq \epsilon s \int_0^\delta e^{-st} \, dt \;\to\; \epsilon \text{ as } s \to \infty. \qquad (1.15.9)$$

Therefore, from (1.15.6),

$$\left| \lim_{s \to \infty} s \int_0^\delta e^{-st} f(x+t)\, dt - f(x) \right| = \left| \lim_{s \to \infty} \left[s \int_0^\delta e^{-st} \left[f(x+t) - f(x) \right] dt \right. \right.$$
$$\left. \left. + s \int_0^\delta e^{-st} f(x)\, dt - f(x) \right] \right|$$
$$\leq \epsilon.$$

Since ϵ was arbitrary, the result follows. □

1.15.2 Laplace Boundedness

If in (1.15.1) the condition $o(1)$ is replaced by $O(1)$ as $s \to \infty$, then f is said to be *right Laplace bounded*, or *Laplace bounded on the right*, at x of order r.

Theorem 1.15.3 *If f is Laplace bounded on the right at x of order r, then f has a right Laplace derivative of order $r-1$ at x and $\overline{LD}_r^+ f(x)$ and $\underline{LD}_r^+ f(x)$ are finite. Conversely, if $\overline{LD}_r^+ f(x)$ and $\underline{LD}_r^+ f(x)$ are finite, then f is Laplace bounded on the right at x of order r.*

□   Let f be Laplace bounded on the right at x of order r. Then

$$s^{r+1} \int_0^\delta e^{-st} \left[f(x+t) - Q(t) \right] dt = O(1) \text{ as } s \to \infty, \tag{1.15.10}$$

where $Q(t)$ is a polynomial of degree at most r. Let $Q(t) = \sum_{i=0}^r b_i t^i$ and define $\overline{Q}(t) = \sum_{i=0}^{r-1} b_i t^i$. From (1.15.9) and Lemma 1.15.1, we have

$$s^r \int_0^\delta e^{-st} \left[f(x+t) - \overline{Q}(t) \right] dt$$
$$= s^r \left[\int_0^\delta e^{-st} \left[f(x+t) - Q(t) \right] dt + \int_0^\delta e^{-st} b_r t^r\, dt \right]$$
$$= \frac{1}{s} O(1) + b_r r! s^{-r-1} + o(1) = o(1) \text{ as } s \to \infty,$$

and, hence, $LD_{r-1}^+ f(x)$ exists finitely and $\overline{Q}(t) = \sum_{i=0}^{r-1} (t^i/i!) LD_i^+ f(x)$. Since $Q(t) = \overline{Q}(t) + b_r t^r$, from (1.15.10) and Lemma 1.15.1,

$$s^{r+1} \int_0^\delta e^{-st} \left[f(x+t) - \sum_{i=0}^{r-1} (t^i/i!) LD_i^+ f(x) \right] dt = O(1) + s^{r+1} \int_0^\delta e^{-st} b_r t^r\, dt$$
$$= O(1) + O(1) = O(1) \text{ as } s \to \infty, \tag{1.15.11}$$

and, so, $\overline{LD}_r^+ f(x)$ and $\underline{LD}_r^+ f(x)$ are finite.

Conversely, suppose that $\overline{LD}_r^+ f(x)$ and $\underline{LD}_r^+ f(x)$ are finite. Then (1.15.11) holds. Let c be any finite constant and let $Q(t) = \sum_{i=0}^{r-1} (t^i/i!) LD_i^+ f(x) + ct^r$. Then, from (1.15.11)

$$
\begin{aligned}
s^{r+1} \int_0^\delta e^{-st} \big[f(x+t) - Q(t) \big]\, dt &= O(1) + s^{r+1} \int_0^\delta e^{-st} ct^r\, dt \\
&= O(1) + O(1) = O(1) \text{ as } s \to \infty.
\end{aligned}
$$

Since $Q(t)$ is a polynomial of degree at most r, f is Laplace bounded on the right of x of order r. □

1.15.3 Bilateral Laplace Derivatives

The *left upper and the left lower Laplace derivates of f at x of order k*, say, written $\overline{LD}_k^- f(x)$ and $\underline{LD}_k^- f(x)$, respectively, are defined as follows. Suppose that the left Laplace derivative of order $r-1$, $LD_{r-1}^- f(x)$, exists finitely and let $Q(t) = \sum_{i=0}^{k-1} (t^i/i!) LD_i^- f(x)$. Define

$$
\overline{LD}_k^- f(x) = \limsup_{s \to \infty} (-1)^k s^{k+1} \int_{-\delta}^0 e^{st} \big[f(x+t) - Q(t) \big]\, dt,
$$

$$
\underline{LD}_k^- f(x) = \liminf_{s \to \infty} (-1)^k s^{k+1} \int_{-\delta}^0 e^{st} \big[f(x+t) - Q(t) \big]\, dt,
$$

respectively. If $\overline{LD}_k^- f(x) = \underline{LD}_k^- f(x)$, the common value is the *left Laplace derivative of f at x of order k* denoted by $LD_k^- f(x)$.

This definition of $LD_k^- f(x)$ is equivalent to the previous definition if it is finite, as we now show.

□ Suppose that $LD_k^- f(x)$ exists in the sense of the previous definition. Then $LD_i^- f(x)$ exists for $i = 0, 1, \cdots, k-1$ and

$$
s^{k+1} \int_{-\delta}^0 e^{st} \Big[f(x+t) - \sum_{i=0}^k (t^i/i!) LD_i^- f(x) \Big]\, dt = o(1) \text{ as } s \to \infty.
$$

So, using Lemma 1.15.1,

$$
\begin{aligned}
s^{k+1} \int_{-\delta}^0 e^{st} \Big[f(x+t) &- \sum_{i=0}^{k-1} (t^i/i!) LD_i^- f(x) \Big]\, dt \\
&= s^{k+1} \int_{-\delta}^0 e^{st} (t^k/k!) LD_k^- f(x) \big]\, dt + o(1) \\
&= s^{k+1} \frac{1}{k!} LD_k^- f(x) \int_0^\delta e^{-st} (-t)^k\, dt + o(1) \\
&= (-1)^k LD_k^- f(x) + o(1).
\end{aligned}
$$

Hence,

$$\lim_{s \to \infty} (-1)^k s^{k+1} \int_{-\delta}^{0} e^{st} \left[f(x+t) - \sum_{i=0}^{k-1} (t^i/i!) LD_i^- f(x) \right] dt = LD_k^- f(x),$$

showing that $LD_k^- f(x)$ is the derivative in the sense of the present definition.

Conversely, if $LD_k^- f(x)$ exists and is finite in the sense of the present definition, then

$$\lim_{s \to \infty} (-1)^k s^{k+1} \int_{-\delta}^{0} e^{st} \left[f(x+t) - \sum_{i=0}^{k-1} (t^i/i!) LD_i^- f(x) \right] dt = LD_k^- f(x),$$

and, so,

$$s^{k+1} \int_{-\delta}^{0} e^{st} \left[f(x+t) - \sum_{i=0}^{k-1} (t^i/i!) LD_i^- f(x) \right] dt = (-1)^k LD_k^- f(x) + o(1) \text{ as } s \to \infty,$$

and using Lemma 1.15.1

$$s^{k+1} \int_{-\delta}^{0} e^{st} \left[f(x+t) - \sum_{i=0}^{k} (t^i/i!) LD_i^- f(x) \right] dt = o(1) \text{ as } s \to \infty,$$

showing that $LD_k^- f(x)$ is the derivative in the sense of the previous definition.
□

Now, if $LD_r^+ f(x)$ and $LD_r^- f(x)$ exist and $LD_i^+ f(x) = LD_i^- f(x)$, $i = 1, \ldots, r$, then we say that f has a *Laplace derivative at x of order r* and this common value is denoted by $LD_r f(x)$, the *(bilateral) Laplace derivative at x of order r*. The *upper and lower (bilateral) Laplace derivates of f at x of order r* are defined by

$$\overline{LD}_r f(x) = \max\{\overline{LD}_r^+ f(x), \overline{LD}_r^- f(x)\},$$
$$\underline{LD}_r f(x) = \min\{\underline{LD}_r^+ f(x), \underline{LD}_r^- f(x)\},$$

respectively, and f is said to be *Laplace bounded of order r at x* if f is both left and right Laplace bounded of order r at x. Analogues of Theorems 15.2 and 15.3 hold.

1.16 Symmetric Laplace Derivatives and Laplace Smoothness

1.16.1 Symmetric Laplace Derivatives and Symmetric Laplace Continuity

Let f be special Denjoy integrable in some neighbourhood $N(x)$ of x and let r be a fixed positive integer. If there exists a polynomial $P(t) = P_x(t)$ of degree at most r such that

$$s^{r+1} \int_0^\delta e^{-st} \left[\frac{f(x+t) + (-1)^r f(x-t)}{2} - \frac{P(t) + (-1)^r P(-t)}{2} \right] dt$$
$$= o(1) \text{ as } s \to \infty \qquad (1.16.1)$$

for some $\delta > 0$ such that $x \pm \delta \in N(x)$, then f is said to have a *symmetric Laplace derivative* at x of order r, and, if $a_r/r!$ is the coefficient of t^r in $P(t)$, then a_r is called the *symmetric Laplace derivative of f at x of order r* and is denoted by $SLD_r f(x)$.

Now, if $0 < \delta_1 < \delta_2$, then, as in the case of the Laplace derivatives, it can be shown that

$$s^{r+1} \int_{\delta_1}^{\delta_2} e^{-st} \left[\frac{f(x+t) + (-1)^r f(x-t)}{2} - \frac{P(t) + (-1)^r P(-t)}{2} \right] dt$$
$$= o(1) \text{ as } s \to \infty$$

and, so, if (1.16.1) is true for $\delta = \delta_1$, it is also true for $\delta = \delta_2$ and, so, (1.16.1) does not depend on δ.

Theorem 1.16.1 *If the derivative $SLD_r f(x)$ exists, it is unique.*

☐   Let $P_1(t) = \sum_{i=0}^r c_i t^i$ and $P_2(t) = \sum_{i=0}^r d_i t^i$ be two polynomials for which (1.16.1) is true. Then, by Lemma 1.15.1

$$s^{r+1} \int_0^\delta e^{-st} \left[P_1(t) - P_2(t) \right] dt = \sum_{i=0}^r \left[(c_i - d_i) s^{r+1} \int_0^\delta e^{-st} t^i \, dt \right]$$
$$= \sum_{i=0}^r \left[(c_i - d_i) i! s^{r-i} \right] + o(1) \text{ as } s \to \infty,$$

$$(1.16.2)$$

and

$$s^{r+1} \int_0^\delta e^{-st} \left[P_1(-t) - P_2(-t) \right] dt = \sum_{i=0}^r \left[(c_i - d_i) s^{r+1} (-1)^i \int_0^\delta e^{-st} t^i \, dt \right]$$
$$= \sum_{i=0}^r \left[(c_i - d_i)(-1)^i i! s^{r-i} \right] + o(1)$$

$$\text{as } s \to \infty.$$

$$(1.16.3)$$

Also, by (1.16.1)

$$\frac{s^{r+1}}{2} \int_0^\delta e^{-st} \left[(P_1(t) - P_2(t)) + (-1)^r (P_1(-t) - P_2(-t)) \right] dt$$

$$= s^{r+1} \int_0^\delta e^{-st} \left[\frac{f(x+t) + (-1)^r f(x-t)}{2} - \frac{P_2(t) + (-1)^r P_2(-t)}{2} \right] dt$$

$$- s^{r+1} \int_0^\delta e^{-st} \left[\frac{f(x+t) + (-1)^r f(x-t)}{2} - \frac{P_1(t) + (-1)^r P_1(-t)}{2} \right] dt$$

$$= o(1) - o(1) = o(1) \text{ as } s \to \infty. \tag{1.16.4}$$

From (1.16.2), (1.16.3) and (1.16.4),

$$\sum_{i=0}^r \left[(c_i - d_i) \frac{(1 + (-1)^{r-i})}{2} i! s^{r-i} \right] = o(1) \text{ as } s \to \infty. \tag{1.16.5}$$

If r is even, $r = 2m$ say, then (1.16.5) is

$$(c_0 - d_0)s^{2m} + (c_2 - d_2)2! s^{2m-2} + \cdots$$
$$\cdots + (c_{2m-2} - d_{2m-2})(2m - 2)! s^2 + (c_{2m} - d_{2m})(2m)! = o(1) \text{ as } s \to \infty. \tag{1.16.6}$$

and so dividing by s^{2m} and then letting $s \to \infty$, we have $c_0 = d_0$ and (1.16.6) becomes

$$(c_2 - d_2)2! s^{2m-2} + \cdots + (c_{2m-2} - d_{2m-2})(2m - 2)! s^2 + (c_{2m} - d_{2m})(2m)!$$
$$= o(1) \text{ as } s \to \infty.$$

Dividing by s^{2m-2} and then letting $s \to \infty$, we have $c_2 = d_2$. Continuing this process we get $c_{2m-2} = d_{2m-2}$ and $(c_{2m} - d_{2m})(2m)! = o(1)$ as $s \to \infty$, and so finally letting $s \to \infty$, $c_{2m} = d_{2m}$.

If r is odd, $r = 2m + 1$ say, then (1.16.5) is

$$(c_1 - d_1)s^{2m} + (c_3 - d_3)3! s^{2m-2} + \cdots + (c_{2m+1} - d_{2m+1})(2m+1)! = o(1) \text{ as } s \to \infty.$$

and, proceeding as above, we have $c_1 = d_1, c_3 = d_3, \ldots, c_{2m+1} = d_{2m+1}$.
    This shows the uniqueness of $SLD_r f(x)$. $\qquad\square$

Theorem 1.16.2 *If $r \geq 2$ and $SLD_r f(x)$ exists, then $SLD_{r-2} f(x)$ exists.*

$\square$ Suppose that $SLD_r f(x)$ exists. Then there is a polynomial $P(t)$ of degree at most r such that (1.16.1) holds. Let $a_j/j!$ be the coefficient of t^j in $P(t)$ and let $\overline{P}(t) = P(t) - \frac{a_{r-1}}{(r-1)!} t^{r-1} - \frac{a_r}{r!} t^r$. Then

$$\overline{P}(t) + (-1)^{r-2}\overline{P}(-t) = P(t) + (-1)^r P(-t) - 2\frac{a_r}{r!} t^r$$

and so, by (1.16.1) and Lemma 1.15.1,

$$s^{r-1} \int_0^\delta e^{-st} \left[\frac{f(x+t) + (-1)^{r-2} f(x-t)}{2} - \frac{\overline{P}(t) + (-1)^{r-2}\overline{P}(-t)}{2} \right] dt$$

$$= s^{r-1} \int_0^\delta e^{-st} \left[\frac{f(x+t) + (-1)^r f(x-t)}{2} - \frac{P(t) + (-1)^r P(-t)}{2} \right] dt$$

$$+ s^{r-1} \int_0^\delta e^{-st} \frac{a_r}{r!} t^r \, dt$$

$$= \frac{1}{s^2} o(1) + \frac{a_r}{r!} r! s^{r-1-r-1} + o(1) = o(1) \text{ as } s \to \infty. \qquad (1.16.7)$$

Since $\overline{P}(t)$ is a polynomial of degree at most $r - 2$, $SLD_{r-2}f(x)$ exists and with value a_{r-2}. ☐

If r is even, then by the above $SLD_2 f(x)$ exists and, as in (1.16.7),

$$s \int_0^\delta e^{-st} \left[\frac{f(x+t) + f(x-t)}{2} - a_0 \right] dt = o(1) \text{ as } s \to \infty.$$

Since $s \int_0^\delta e^{-st} \, dt \to 1$ as $s \to \infty$, $SLD_0 f(x)$ exists and

$$SLD_0 f(x) = a_0 = \lim_{s \to \infty} s \int_0^\delta e^{-st} \frac{f(x+t) + f(x-t)}{2} \, dt.$$

Similarly, if r is odd, $SLD_1 f(x)$ exists.

Theorem 1.16.3 (a) *If* $\lim_{t \to 0} \frac{1}{2} [f(x+t) + f(x-t)]$ *exists finitely, with value* ℓ_0, *say, then* $SLD_0 f(x)$ *exists and* $SLD_0 f(x) = \ell_0$.
(b) *If* $\lim_{t \to 0} \frac{f(x+t) - f(x-t)}{2t}$ *exists finitely, with value* ℓ_1, *say, then* $SLD_1 f(x)$ *exists and* $SLD_1 f(x) = \ell_1$.

☐   (a) Let ℓ_0 be as defined in (a) above. Then for arbitrary $\epsilon > 0, \delta > 0$ there is a σ, $0 < \sigma < \delta$, such that $\left| \frac{f(x+t) + f(x-t)}{2} - \ell_0 \right| < \epsilon$ for $0 < t < \sigma$, and so

$$s \int_0^\sigma e^{-st} \left[\frac{f(x+t) + f(x-t)}{2} - \ell_0 \right] dt < \epsilon s \int_0^\sigma e^{-st} \, dt.$$

Hence,

$$s \int_0^\delta e^{-st} \left[\frac{f(x+t) + f(x-t)}{2} - \ell_0 \right] dt$$

$$= \left(s \int_0^\sigma + s \int_\sigma^\delta \right) e^{-st} \left[\frac{f(x+t) + f(x-t)}{2} - \ell_0 \right] dt$$

$$\leq \epsilon s \int_0^\sigma e^{-st} \, dt + s \int_\sigma^\delta e^{-st} \left[\frac{f(x+t) + f(x-t)}{2} - \ell_0 \right] dt.$$

Letting $s \to \infty$,

$$\limsup_{s \to \infty} s \int_0^\delta e^{-st} \left[\frac{f(x+t) + f(x-t)}{2} - \ell_0 \right] dt \le \epsilon.$$

Similarly,

$$\liminf_{s \to \infty} s \int_0^\delta e^{-st} \left[\frac{f(x+t) + f(x-t)}{2} - \ell_0 \right] dt \ge -\epsilon.$$

Since $\epsilon > 0$ is arbitrary,

$$\lim_{s \to \infty} s \int_0^\delta e^{-st} \left[\frac{f(x+t) + f(x-t)}{2} - \ell_0 \right] dt \le \epsilon,$$

showing that $SLD_0 f(x) = \ell_0$.

(b) Let ℓ_1 be as defined in (b) above. Then for arbitrary $\epsilon > 0, \delta > 0$ there is a $\sigma, 0 < \sigma < \delta$, such that $\left| \dfrac{f(x+t) - f(x-t)}{2t} - \ell_1 \right| < \epsilon$ for $0 < t < \sigma$, or equivalently $\left| \dfrac{f(x+t) - f(x-t)}{2} - \ell_1 t \right| < \epsilon t$ for $0 < t < \sigma$. So, by Lemma 1.15.1,

$$s \int_0^\sigma e^{-st} \left[\frac{f(x+t) - f(x-t)}{2} - \ell_1 t \right] dt \le \epsilon s \int_0^\sigma e^{-st} t \, dt$$

$$= \frac{\epsilon}{s} + o(1) \text{ as } s \to \infty.$$

So,

$$s \int_0^\delta e^{-st} \left[\frac{f(x+t) - f(x-t)}{2} - \ell_1 t \right] dt$$

$$= \left(s \int_o^\sigma + s \int_\sigma^\delta \right) e^{-st} \left[\frac{f(x+t) - f(x-t)}{2} - \ell_1 t \right] dt$$

$$\le \frac{\epsilon}{s} + s \int_\sigma^\delta e^{-st} \left[\frac{f(x+t) - f(x-t)}{2} - \ell_1 t \right] dt + o(1).$$

Hence,

$$\limsup_{s \to \infty} s \int_0^\delta e^{-st} \left[\frac{f(x+t) - f(x-t)}{2} - \ell_1 t \right] dt \le 0.$$

Similarly,

$$\liminf_{s \to \infty} s \int_0^\delta e^{-st} \left[\frac{f(x+t) - f(x-t)}{2} - \ell_1 t \right] dt \ge 0,$$

and, so,

$$\lim_{s \to \infty} s \int_0^\delta e^{-st} \left[\frac{f(x+t) - f(x-t)}{2} - \ell_1 t \right] dt = 0$$

and, hence, $SLD_1 f(x) = \ell_1$. $\qquad \square$

Definition. A special Denjoy integrable function f is said to be *symmetric Laplace continuous at x of even order* if

$$\lim_{s \to \infty} s \int_0^\delta e^{-st} \frac{f(x+t) + f(x-t)}{2} \, dt = f(x),$$

and f is said to be *symmetric Laplace continuous at x of odd order* if

$$\lim_{s \to \infty} s \int_0^\delta e^{-st} \left[f(x+t) - f(x-t) \right] dt = 0.$$

It can be verified that if f is special Denjoy integrable in some neighbourhood of x, then f is symmetric Laplace continuous at x of even order if f is symmetric continuous of even order at x, and f is symmetric Laplace continuous at x of odd order if f is symmetric continuous of odd order at x.

To define the *upper and lower symmetric Laplace derivates of f at x of order k* we suppose that $SLD_{k-2} f(x)$ exists. If k is even, $k = 2m$ say, define

$$\overline{SLD}_{2m} f(x)$$

$$= \limsup_{s \to \infty} s^{2m+1} \int_0^\delta e^{-st} \left[\frac{f(x+t) + f(x-t)}{2} - \sum_{i=0}^{m-1} \frac{t^{2i}}{(2i)!} SLD_{2i} f(x) \right] dt,$$

with a similar definition, but taking lim inf, for $\underline{SLD}_{2m} f(x)$.

If k is odd, $k = 2m + 1$, say, define

$$\overline{SLD}_1 f(x) = \limsup_{s \to \infty} s^2 \int_0^\delta e^{-st} \frac{f(x+t) - f(x-t)}{2} \, dt, \text{ if } m = 0;$$

and if $m \geq 1$,

$$\overline{SLD}_{2m+1} f(x) =$$

$$\limsup_{s \to \infty} s^{2m+2} \int_0^\delta e^{-st} \left[\frac{f(x+t) - f(x-t)}{2} - \sum_{i=0}^{m-1} \frac{t^{2i+1}}{(2i+1)!} SLD_{2i+1} f(x) \right] dt,$$

with a similar definition, but taking lim inf, for $\underline{SLD}_{2m+1} f(x)$.

If $\overline{SLD}_k f(x) = \underline{SLD}_k f(x)$, then the common value is the *symmetric Laplace derivative of f at x of order k*, possibly infinite, $SLD_k f(x)$. Clearly this definition agrees with the previous one if $SLD_k f(x)$ is finite.

1.16.2 Symmetric Laplace Boundedness

If, in (1.16.1), $o(1)$ is replaced by $O(1)$ as $s \to \infty$, then f is said to be *symmetric Laplace bounded at x of order r.*

Theorem 1.16.4 *If f is symmetric Laplace bounded at x of order r, then f has a finite symmetric Laplace derivative at x of order $r - 2$ and $\overline{SLD}_r f(x)$ and $\underline{SLD}_r f(x)$ are finite. Conversely, if $\overline{SLD}_r f(x)$ and $\underline{SLD}_r f(x)$ are finite, then f is symmetric Laplace bounded at x of order r*

☐   Let f be symmetric Laplace bounded at x of order r. Then there is a polynomial $P(t)$ of degree at most r such that

$$s^{r+1}\int_0^\delta e^{-st}\left[\frac{f(x+t)+(-1)^r f(x-t)}{2}-\frac{P(t)+(-1)^r P(-t)}{2}\right]dt = O(1)$$

$$\text{as } s\to\infty.$$

Let $P(t)=\sum_{i=0}^r b_i t^i$. Then writing $\overline{P}(t)=\sum_{i=0}^{r-2}b_i t^i$ we have

$$s^{r-1}\int_0^\delta e^{-st}\left[\frac{f(x+t)+(-1)^{r-2}f(x-t)}{2}-\frac{\overline{P}(t)+(-1)^{r-2}\overline{P}(-t)}{2}\right]dt$$

$$= s^{r-1}\int_0^\delta e^{-st}\left[\frac{f(x+t)+(-1)^r f(x-t)}{2}-\frac{P(t)+(-1)^r P(-t)}{2}\right]dt$$

$$+s^{r-1}\int_0^\delta e^{-st}b_r t^r\,dt$$

$$= \frac{1}{s^2}O(1)+b_r r!s^{r-1-r-1}+o(1)=o(1)\text{ as }s\to\infty. \tag{1.16.8}$$

Since $\overline{P}(t)$ is a polynomial of degree at most $r-2$, $SLD_{r-2}f(x)$ exists. Further,

$$\overline{P}(t)+(-1)^{r-2}\overline{P}(-t) = \begin{cases} \sum_{i=0}^{m-1}\frac{t^{2i}}{(2i)!}SLD_{2i}f(x), & \text{if } r=2m, \\[2mm] \sum_{i=0}^{m-1}\frac{t^{2i+1}}{(2i+1)!}SLD_{2i+1}f(x), & \text{if } r=2m+1. \end{cases}$$

So, from (1.16.9),

$$s^{r+1}\int_0^\delta e^{-st}\left[\frac{f(x+t)+(-1)^{r-2}f(x-t)}{2}-\frac{\overline{P}(t)+(-1)^{r-2}\overline{P}(-t)}{2}\right]dt$$

$$= s^{r+1}\int_0^\delta e^{-st}\left[\frac{f(x+t)+(-1)^r f(x-t)}{2}-\frac{P(t)+(-1)^r P(-t)}{2}\right]dt$$

$$+s^{r+1}\int_0^\delta e^{-st}b_r t^r\,dt$$

$$= O(1)+b_r r!s^{r+1-r-1}+o(1)\text{ as }s\to\infty. \tag{1.16.9}$$

Hence, taking lim sup and lim inf in (1.16.10), $\overline{SLD}_r f(x)$ and $\underline{SLD}_r f(x)$ are finite.

Conversely, let $\overline{SLD}_r f(x)$ and $\underline{SLD}_r f(x)$ be finite. Then from the definition it follows that $SLD_{r-2}f(x)$ exists finitely and so for a polynomial $P(t)$ of degree at most $r-2$

$$s^{r+1}\int_0^\delta e^{-st}\left[\frac{f(x+t)+(-1)^r f(x-t)}{2}-\frac{P(t)+(-1)^{r-2}P(-t)}{2}\right]dt$$

$$=O(1)\text{ as }s\to\infty. \tag{1.16.10}$$

Let $P_1(t) = P(t) + ct^{r-1} + dt^r$ where c and d are constants. Then, from (1.16.11) and by Lemma 1.15.1,

$$s^{r+1} \int_0^\delta e^{-st}\left[\frac{f(x+t) + (-1)^r f(x-t)}{2} - \frac{P_1(t) + (-1)^r P_1(-t)}{2}\right] dt$$

$$= s^{r+1} \int_0^\delta e^{-st}\left[\frac{f(x+t) + (-1)^r f(x-t)}{2} - \frac{P(t) + (-1)^{r-2} P(-t)}{2}\right] dt$$

$$- s^{r+1} \int_0^\delta e^{-st} dt^r\, dt$$

$$= O(1) - dr! s^{r+1-r-1} - o(1) = O(1) \text{ as } s \to \infty,$$

and so f is symmetric Laplace bounded at x of order r.   □

1.16.3   Laplace Smoothness

Let f be special Denjoy integrable in some neighbourhood of x and let $SLD_{r-2}f(x)$ exist finitely. Let

$$T(t) = \sum_{i=1}^{} \frac{t^{r-2i}}{(r-2i)!} SLD_{r-2i}f(x), \qquad (1.16.11)$$

where the summation in (1.16.12) extends to $i = r/2$ if r is even and to $i = (r-1)/2$ if r is odd. Let

$$\overline{LS}_r f(x) = \limsup_{s\to\infty} s^r \int_0^\delta e^{-st}\left[\frac{f(x+t) + (-1)^r f(x-t)}{2} - T(t)\right] dt,$$

$$\underline{LS}_r f(x) = \liminf_{s\to\infty} s^r \int_0^\delta e^{-st}\left[\frac{f(x+t) + (-1)^r f(x-t)}{2} - T(t)\right] dt.$$

Then $\overline{LS}_r f(x)$ and $\underline{LS}_r f(x)$ are called the *upper* and *lower index of Laplace smoothness* of f of order r at x. If $\overline{LS}_r f(x) = \underline{LS}_r f(x) = LS_r f(x) = 0$, then f is said to be *Laplace smooth of order r at x*.

Theorem 1.16.5 *If $\overline{SLD}_r f(x)$ and $\underline{SLD}_r f(x)$ are finite, then f is Laplace smooth of order r at x.*

□   Since $\overline{SLD}_r f(x)$ and $\underline{SLD}_r f(x)$ are finite, by definition $SLD_{r-2}f(x)$ exists and

$$s^{r+1} \int_0^\delta e^{-st}\left[\frac{f(x+t) + (-1)^r f(x-t)}{2} - T(t)\right] dt = O(1) \text{ as } s \to \infty,$$

where $T(t)$ is as defined in (1.16.12). So,

$$s^r \int_0^\delta e^{-st}\left[\frac{f(x+t) + (-1)^r f(x-t)}{2} - T(t)\right] dt = o(1) \text{ as } s \to \infty,$$

completing the proof.   □

Chapter 2

Relations between Derivatives

In this chapter, we shall study relations between the various derivatives considered in Chapter I where all the appropriate notations are defined. All the kth order derivatives considered are generalizations of the kth order ordinary derivative $f^{(k)}$. We first show that this is the case for the kth order Peano derivative $f_{(k)}$.

2.1 Ordinary and Peano Derivatives, $f^{(k)}$ and $f_{(k)}$

The following theorem sharpens Theorem 1.4.1 of Chapter I.

Theorem 2.1.1 If $f^{(k)}(x)$ exists finitely so does $f_{(k)}(x)$ and $f_{(k)}(x) = f^{(k)}(x)$. Moreover, if $f^{(k)}(x)$ exists finitely

$$D_+ f^{(k)}(x) \leq \underline{f}^+_{(k+1)}(x) \leq \overline{f}^+_{(k+1)}(x) \leq D^+ f^{(k)}(x),$$

where $D_+ f^{(k)} = D_+(f^{(k)})$ and $D^+ f^{(k)} = D^+(f^{(k)})$ are, respectively, the lower and upper Dini right derivates of $f^{(k)}$. Similarly,

$$D_- f^{(k)}(x) \leq \underline{f}^-_{(k+1)}(x) \leq \overline{f}^-_{(k+1)}(x) \leq D^- f^{(k)}(x).$$

☐   The proof of the first part is in Theorem 1.4.1 of Chapter I. For the second part, suppose that $f^{(k)}(x)$ exists finitely. Then so does $f_{(k)}(x)$ and so let

$$\gamma_{k+1}(f; x, t) = \frac{(k+1)!}{t^{k+1}} \left[f(x+t) - \sum_{i=0}^{k} \frac{t^i}{i!} f_{(i)}(x) \right].$$

Then, as in the relation (1.4.5) of Chapter I, since $f_{(i)}(x) = f^{(i)}(x)$ for $i = 0, 1, \ldots, k$,

$$\gamma_{k+1}(f; x, t) = \frac{f^{(k)}(x + \xi_t) - f^{(k)}(x)}{\xi_t}, \quad 0 < \xi_t < t.$$

Letting $t \to 0+$,

$$\overline{f}^+_{(k+1)}(x) = \limsup_{t \to 0+} \gamma_{k+1}(f; x, t) = \limsup_{t \to 0+} \frac{f^{(k)}(x + \xi_t) - f^{(k)}(x)}{\xi_t}$$

$$\leq \limsup_{t \to 0+} \frac{f^{(k)}(x + t) - f^{(k)}(x)}{t} = D^+ f^{(k)}(x).$$

This proves the right inequality in the first set of inequalities and the rest of the inequalities can be proved in a similar manner. □

Since

$$\overline{f}_{(k+1)}(x) = \max\{\overline{f}^+_{(k+1)}(x), \overline{f}^-_{(k+1)}(x)\}, \ \underline{f}_{(k+1)}(x) = \min\{\underline{f}^+_{(k+1)}(x), \underline{f}^-_{(k+1)}(x)\}$$

and

$$\overline{D}f^{(k)}(x) = \max\{\overline{D}^+ f^{(k)}(x), \overline{D}^- f^{(k)}(x)\}, \ \underline{D}f^{(k)}(x) = \min\{\underline{D}^+ f^{(k)}(x), \underline{D}^- f^{(k)}(x)$$

we have from the above

$$\underline{D}f^{(k)}(x) \leq \underline{f}_{(k+1)}(x) \leq \overline{f}_{(k+1)}(x) \leq \overline{D}f^{(k)}(x).$$

The converse of Theorem 2.1.1 is not true; see the example following Theorem 1.4.1 of Chapter I.

2.2   Riemann* and Peano Derivatives, $f^*_{(k)}$ and $f_{(k)}$

Theorem 2.2.1 *If $k \geq 2$ and if $f^*_{(k-1)}(x_0)$ exists finitely, then $f_{(k-1)}(x_0)$ also exists finitely and $f_{(i)}(x_0) = f^*_{(i)}(x_0)$ for $i = 1, 2, \ldots, k - 1$. Moreover,*

$$\lim_{x_{k-1} \to x_0} \cdots \lim_{x_1 \to x_0} k! Q_k(f; x_0, \ldots, x_k) = \gamma_k(f; x_0, x_k - x_0) \qquad (2.2.1)$$

and, hence, $f^*_{(k)}(x_0)$ exists, possibly infinitely, if and only if $f_{(k)}(x_0)$ exists; in either case, $f^*_{(k)}(x_0) = f_{(k)}(x_0)$.

□   Since $f^*_{(k-1)}(x_0)$ exists finitely, by Theorem 1.5.2 of Chapter I, $f^*_{(i)}(x_0)$ exists finitely for $i = 1, 2, \ldots, k - 1$ and so

$$\lim_{x_i \to x_0} \cdots \lim_{x_1 \to x_0} i! Q_i(f; x_0, \ldots, x_i) = f^*_{(i)}(x_0), \ i = 1, 2, \ldots, k - 1. \qquad (2.2.2)$$

By Theorem 1.5.1 of Chapter I, f is continuous at x_0 and so $\lim_{x_1 \to x_0} Q_1 f: x_1, x_2 = Q_1(f; x_0, x_2)$. Also, since by (2.2.2) we have $\lim_{x_1 \to x_0} Q_1 f: x_0, x_1 =$

$f_{(1)}^*(x_0) = f'(x_0)$, it follows that

$$
\begin{aligned}
\lim_{x_1 \to x_0} Q_2(f; x_0, x_1, x_2) &= \lim_{x_1 \to x_0} \frac{Q_1(f; x_0, x_1) - Q_1(f; x_1, x_2)}{x_0 - x_2} \\
&= \frac{f'(x_0) - Q_1(f; x_0, x_2)}{x_0 - x_2} \\
&= \frac{f(x_2) - f(x_0) - (x_2 - x_0)f'(x_0)}{(x_2 - x_0)^2} \\
&= \frac{1}{2!} \gamma_2(f; x_0, x_2 - x_0),
\end{aligned}
$$

proving the result in the case $k = 2$.

Suppose that the result is true for $k = r$ and that $f_{(r)}^*(x_0)$ exists finitely. Then by Theorem 1.5.2 of Chapter I, $f_{(i)}^*(x_0)$ exists finitely for $i = 1, 2, \ldots, r$, and (2.2.2) is true for $i = 1, 2, \ldots, r$. Since the result is assumed for $k = r$, $f_{(r-1)}(x_0)$ exists finitely and $f_{(i)}^*(x_0) = f_{(i)}(x_0)$ for $i = 1, 2, \ldots, r - 1$, and by (2.2.1)

$$
\lim_{x_{r-1} \to x_0} \cdots \lim_{x_1 \to x_0} r! Q_r(f; x_0, \ldots, x_r) = \gamma_r(f; x_0, x_r - x_0). \tag{2.2.3}
$$

Since $f_{(r)}^*(x_0)$ exists, taking the limit in (2.2.3) as $x_r \to x_0$,

$$
\lim_{x_r \to x_0} \cdots \lim_{x_1 \to x_0} r! Q_r(f; x_0, x_1, \ldots, x_r) = f_{(r)}(x_0). \tag{2.2.4}
$$

Since f is continuous at x_0,

$$
\lim_{x_1 \to x_0} Q_r(f; x_1, x_2, \ldots, x_{r+1}) = Q_r(f; x_0, x_2, \ldots, x_{r+1}). \tag{2.2.5}
$$

So, from (2.2.5) and (2.2.3),

$$
\begin{aligned}
&\lim_{x_r \to x_0} \cdots \lim_{x_1 \to x_0} r! Q_r(f; x_1, x_2 \ldots, x_{r+1}) \\
&= \lim_{x_r \to x_0} \cdots \lim_{x_2 \to x_0} r! Q_r(f; x_0, x_2 \ldots, x_{r+1}) \\
&= \gamma_r(f; x_0, x_{r+1} - x_0). \tag{2.2.6}
\end{aligned}
$$

From (2.2.4) and (2.2.6),

$$
\begin{aligned}
&\lim_{x_r \to x_0} \cdots \lim_{x_1 \to x_0} (r+1)! Q_{r+1}(f; x_0, x_1 \ldots, x_{r+1}) \\
&= \frac{r+1}{x_0 - x_{r+1}} \lim_{x_r \to x_0} \cdots \\
&\quad \times \lim_{x_1 \to x_0} \left[r! Q_r(f; x_0, x_1, \ldots, x_r) - r! Q_r(f; x_1, x_2 \ldots, x_{r+1}) \right] \\
&= \frac{r+1}{x_0 - x_{r+1}} \left[f_{(r)}(x_0) - \gamma_r(f; x_0, x_{r+1} - x_0) \right] = \gamma_{r+1}(f; x_0, x_{r+1} - x_0).
\end{aligned}
$$

$$\tag{2.2.7}$$

The relation (2.2.7) shows that (2.2.1) is true for $k = r + 1$. So, the result is proved by induction. $\qquad\square$

Theorem 2.2.2 *If $k \geq 2$, f is continuous at x_0 and $f_{(k-1)}(x_0)$ exists finitely, then $f^*_{(k-1)}(x_0)$ exists finitely and $f^*_{(i)}(x_0) = f_{(i)}(x_0)$ for $i = 1, 2, \ldots, k-1$. Moreover the relation (2.2.1) holds. Hence, $f_{(k)}(x_0)$ exists, possibly infinitely, if and only if $f^*_{(k)}(x_0)$ exists and in either case $f^*_{(k)}(x_0) = f_{(k)}(x_0)$.*

$\square$ Let $k = 2$; f is continuous at x_0 so $\lim_{x_1 \to x_0} Q_1(f; x_1, x_2) = Q_1(f; x_0, x_2)$; also, $\lim_{x_1 \to x_0} Q_1(f; x_0, x_1) = f'(x_0) = f_{(1)}(x_0)$. Hence, as in Theorem 2.2.1,

$$\lim_{x_1 \to x_0} 2! Q_2(f; x_0, x_1, x_2) = \gamma_2(f; x_0, x_2 - x_0)$$

completing the proof for the case $k = 2$.

Suppose that the result is true for $k = r$ and that $f_{(r)}(x_0)$ exists finitely. Then $f_{(r-1)}(x_0)$ exists finitely and since the result is true for $k = r$, $f^*_{(i)}(x_0)$ exists finitely and $f^*_{(i)}(x_0) = f_{(i)}(x_0)$ for $i = 1, 2, \ldots, r-1$, and the relation (2.2.3) holds. Since $f_{(r)}(x_0)$ exists taking the limit in (2.2.3) as $x_r \to 0$, (2.2.4) holds and $f^*_{(r)}(x_0)$ exists with $f^*_{(r)}(x_0) = f_{(r)}(x_0)$. Since f is continuous at x_0 and the result is true for $k = r$, (2.2.6) holds. Hence, as in the previous theorem, we obtain (2.2.7) showing that (2.2.1) is also true for $k = r + 1$ and completing the proof. $\qquad\square$

Remark. It follows from Theorems 1.1 and 1.2 that $f^*_{(k)}$ and $f_{(k)}$ are the same in the sense that if one of them exists, so does the other with equal value and, further, the relation (2.2.1) shows that the upper, lower and unilateral derivates are the same.

2.3 Symmetric Riemann* and Symmetric de la Vallée Poussin Derivatives, $f^{*(s)}_{(k)}$ and $f^{(s)}_{(k)}$

Theorem 2.3.1 *Let $k \geq 2$.*
(a) If $f^{(s)}_{(2k-2)}(x_0)$ exists finitely, then

$$\lim_{h_{k-1} \to 0} \cdots \lim_{h_1 \to 0} (2k)! Q_{2k}(f; x_0 - h_k, \ldots, x_0 - h_1, x_0, x_0 + h_1, \ldots, x_0 + h_k)$$
$$= \varpi_{2k}(f; x_0, h_k) \qquad (2.3.1)$$

and so $f^{(s)}_{(2k)}(x_0)$ exists if and only if $f^{*(s)}_{(2k)}(x_0)$ exists and in either case they are equal;

(b) if $f_{(2k-3)}^{(s)}(x_0)$ exists finitely, then

$$\lim_{h_{k-1}\to 0} \cdots \lim_{h_1\to 0} (2k-1)! Q_{2k-1}(f; x_0 - h_k, \ldots, x_0 - h_1, x_0 + h_1, \ldots, x_0 + h_k)$$

$$= \varpi_{2k-1}(f; x_0, h_k), \qquad (2.3.2)$$

and so $f_{(2k-1)}^{(s)}(x_0)$ exists if and only if $f_{(2k-1)}^{*(s)}(x_0)$ exists and, in either case, they are equal.

□   We prove (a) and discuss briefly the proof of (b).

We may suppose without loss in generality that $x_0 = 0$ and, by adding a suitable constant if necessary, that $f(x_0) = 0$.

Applying (1.1.1) of Chapter I,

$$Q_{2k}(f; -h_k, \ldots, -h_1, 0, h_1, \ldots, h_k) =$$
$$\frac{Q_{2k-1}(f; -h_k, \ldots, -h_1, 0, h_1, \ldots, h_{k-1}) - Q_{2k-1}(f; -h_{k-1}, \ldots, -h_1, 0, h_1, \ldots, h_k)}{-2h_k}$$

$$= \frac{1}{2h_k(h_{k-1} + h_k)} \Bigg[Q_{2k-2}(f; -h_k, \ldots, -h_1, 0, h_1, \ldots, h_{k-2})$$

$$- Q_{2k-2}(f; -h_{k-1}, \ldots - h_1, 0, h_1, \ldots, h_{k-1})$$

$$- Q_{2k-2}(f; -h_{k-1}, \ldots, -h_1, 0, h_1, \ldots, h_{k-1})$$

$$+ Q_{2k-2}(f; -h_{k-2}, \ldots, -h_1, 0, h_1, \ldots, h_k) \Bigg]$$

$$= \frac{1}{2h_k(h_{k-1} + h_k)} \Bigg[Q_{2k-2}(f; -h_k, \ldots, -h_1, 0, h_1, \ldots, h_{k-2})$$

$$+ Q_{2k-2}(f; -h_{k-2}, \ldots, -h_1, 0, h_1, \ldots, h_k)$$

$$- 2Q_{2k-2}(f; -h_{k-1}, \ldots - h_1, 0, h_1, \ldots, h_{k-1}) \Bigg].$$

$$(2.3.3)$$

Assume now that $k = 2$. Then (2.3.3) reduces to

$$Q_4(f; -h_2, -h_1, 0, h_1, h_2)$$
$$= \frac{Q_2(f; -h_2, -h_1, 0) + Q_2(f; 0, h_2, h_1) - 2Q_2(f; -h_1, 0, h_1)}{2h_2(h_1 + h_2)}. \qquad (2.3.4)$$

Since $f(0) = 0$, we have the following:

$$Q_2(f; -h_2, -h_1, 0) = \frac{f(-h_2)}{h_2(h_2 - h_1)} - \frac{f(-h_1)}{h_1(h_2 - h_1)}, \qquad (2.3.5)$$

$$Q_2(f; 0, h_2, h_1) = \frac{f(h_2)}{h_2(h_2 - h_1)} - \frac{f(h_1)}{h_1(h_2 - h_1)}, \qquad (2.3.6)$$

$$Q_2(f; -h_1, 0, h_1) = \frac{f(-h_1)}{2h_1^2} - \frac{f(h_1)}{2h_1^2}. \qquad (2.3.7)$$

From (2.3.4), (2.3.5), (2.3.6) and (2.3.7),

$$Q_4(f; -h_2, -h_1, 0, h_1, h_2) = \frac{\dfrac{f(h_2) + f(-h_2)}{h_2(h_2 - h_1)} - \dfrac{f(h_1) + f(-h_1)}{h_1(h_2 - h_1)} - \dfrac{f(h_1) + f(-h_1)}{h_1^2}}{2h_2(h_1 + h_2)}$$

and, hence,

$$\lim_{h_1 \to 0} 4! Q_4(f; -h_2, -h_1, 0, h_1, h_2)$$

$$= \lim_{h_1 \to 0} 4! \frac{\dfrac{f(h_2) + f(-h_2)}{h_2(h_2 - h_1)} - \dfrac{f(h_1) + f(-h_1)}{h_1^2}\left(\dfrac{h_1}{h_2 - h_1} + 1\right)}{2h_2(h_1 + h_2)}$$

$$= \frac{4!}{2h_2^2}\left[\frac{f(h_2) + f(-h_2)}{h_2^2} - f_{(2)}^{(s)}(0)\right]$$

$$= \frac{4!}{h_2^4}\left[\frac{f(h_2) + f(-h_2)}{2} - \frac{h_2^2}{2!}f_{(2)}^{(s)}(0)\right] = \varpi_4(f; 0, h_2),$$

proving the result for $k = 2$.

We now suppose that the result is true for $k = r \geq 2$, and prove it for $k = r + 1$. Let $k = r + 1$ and assume that $f_{(2r)}^{(s)}(0)$ exists finitely. Then, from (2.3.3), we have

$$Q_{2r+2}(f; -h_{r+1}, \ldots, -h_1, 0, h_1, \ldots, h_{r+1}) =$$

$$= \frac{1}{h_k(h_{k-1} + h_k)}\left[\frac{1}{2}\left[Q_{2r}(f; -h_{r+1}, \ldots, -h_1, 0, h_1, \ldots, h_{r-1})\right.\right.$$

$$+ Q_{2r}(f; -h_{r-1}, \ldots, -h_1, 0, h_1, \ldots, h_{r+1})\Big]$$

$$\left. - Q_{2r}(f; -h_r, \ldots - h_1, 0, h_1, \ldots, h_r)\right]. \quad (2.3.8)$$

Since $f(0) = 0$,

$$Q_{2r}^- = Q_{2r}(f; -h_{r+1}, \ldots, -h_1, 0, h_1, \ldots, h_{r-1}) \quad (2.3.9)$$

$$= \sum_{j=1}^{r+1} \frac{f(-h_j)}{h_j \prod_{\substack{i=1 \\ i \neq j}}^{r+1}(h_j - h_i)\prod_{i=1}^{r-1}(h_j + h_i)}$$

$$+ \sum_{j=1}^{r-1} \frac{f(h_j)}{h_j \prod_{\substack{i=1 \\ i \neq j}}^{r-1}(h_j - h_i)\prod_{i=1}^{r+1}(h_j + h_i)},$$

$$Q_{2r}^+ = Q_{2r}(f; -h_{r-1}, \ldots, -h_1, 0, h_1, \ldots, h_{r+1}) \quad (2.3.10)$$

$$= \sum_{j=1}^{r-1} \frac{f(-h_j)}{h_j \prod_{\substack{i=1 \\ i \neq j}}^{r-1}(h_j - h_i)\prod_{i=1}^{r+1}(h_j + h_i)}$$

$$+ \sum_{j=1}^{r+1} \frac{f(h_j)}{h_j \prod_{\substack{i=1 \\ i \neq j}}^{r+1}(h_j - h_i)\prod_{i=1}^{r-1}(h_j + h_i)}.$$

From (2.3.9) and (2.3.10),

$$\frac{Q_{2r}^- + Q_{2r}^+}{2} \tag{2.3.11}$$

$$= \sum_{j=1}^{r+1} \frac{\dfrac{f(h_j) + f(-h_j)}{2}}{h_j \prod_{\substack{i=1 \\ i \neq j}}^{r+1}(h_j - h_i) \prod_{i=1}^{r-1}(h_j + h_1)}$$

$$+ \sum_{j=1}^{r-1} \frac{\dfrac{f(h_j) + f(-h_j)}{2}}{h_j \prod_{\substack{i=1 \\ i \neq j}}^{r-1}(h_j - h_i) \prod_{i=1}^{r+1}(h_j + h_1)}.$$

Since

$$\frac{f(h_j) + f(-h_j)}{2} = \sum_{i=1}^{r} \frac{h_j^{2i}}{(2i)!} f_{(2i)}^s(0) + \frac{h_j^{2r+2}}{(2r + 2)!} \varpi_{2r+2}(f; 0, h_j), \tag{2.3.12}$$

we get from (2.3.11)

$$\frac{Q_{2r}^- + Q_{2r}^+}{2} = I_1 + I_2 + I_3 + I_4 \tag{2.3.13}$$

where

$$I_1 = \sum_{j=1}^{r+1} \frac{\sum_{i=1}^{r} \frac{h_j^{2i}}{(2i)!} f_{(2i)}^s(0)}{h_j \prod_{\substack{i=1 \\ i \neq j}}^{r+1}(h_j - h_i) \prod_{i=1}^{r-1}(h_j + h_1)},$$

$$I_2 = \sum_{j=1}^{r+1} \frac{\frac{h_j^{2r+2}}{(2r+2)!} \varpi_{2r+2}(f; 0, h_j)}{h_j \prod_{\substack{i=1 \\ i \neq j}}^{r+1}(h_j - h_i) \prod_{i=1}^{r-1}(h_j + h_1)},$$

$$I_3 = \sum_{j=1}^{r-1} \frac{\sum_{i=1}^{r} \frac{h_j^{2i}}{(2i)!} f_{(2i)}^s(0)}{h_j \prod_{\substack{i=1 \\ i \neq j}}^{r-1}(h_j - h_i) \prod_{i=1}^{r+1}(h_j + h_1)},$$

$$I_4 = \sum_{j=1}^{r-1} \frac{\frac{h_j^{2r+2}}{(2r+2)!} \varpi_{2r+2}(f; 0, h_j)}{h_j \prod_{\substack{i=1 \\ i \neq j}}^{r-1}(h_j - h_i) \prod_{i=1}^{r+1}(h_j + h_1)}.$$

Let $P(x) = \sum_{i=1}^{r} \frac{x^{2i}}{(2i)!} f_{(2i)}^{(s)}(0)$. Then since $P(0) = 0$ and $P(x) = P(-x)$, $I_1 + I_3$ is the divided difference of $P(x)$ at the points $-h_{r+1}, \ldots, -h_1, 0, h_1, \ldots, h_{r-1}$, and also at the points $-h_{r-1}, \ldots, -h_1, 0, h_1, \ldots, h_{r+1}$. Since $P(x)$ is a polynomial of degree at most $2r$, this divided difference is equal to the coefficient of x^{2r}. Hence,

$$I_1 + I_3 = \frac{1}{(2r)!} f_{(2r)}^{(s)}(0). \tag{2.3.14}$$

To estimate I_2, note that since $f^{(s)}_{(2r)}(0)$ exists

$$\frac{f(h_j) + f(-h_j)}{2} = \sum_{i=1}^{r} \frac{h_j^{2i}}{(2i)!} f^{(s)}_{(2i)}(0) + o(h_j^{2r}), \tag{2.3.15}$$

and so, from (2.3.12) and (2.3.15), $\dfrac{h_j^{2r+2}}{(2r+2)!} \varpi_{2r+2}(f; 0, h_j) = o(h_j^{2r})$. Using this we have from (2.3.13)

$$
\begin{aligned}
I_2 &= \sum_{j=1}^{r-1} \frac{o(h_j^{2r})}{h_j \prod_{\substack{i=1 \\ i \neq j}}^{r+1}(h_j - h_i) \prod_{i=1}^{r-1}(h_j + h_i)} \\
&\quad + \sum_{j=r}^{r+1} \frac{\frac{h_j^{2r+2}}{(2r+2)!}\varpi_{2r+2}(f; 0, h_j)}{h_j \prod_{\substack{i=1 \\ i \neq j}}^{r+1}(h_j - h_i) \prod_{i=1}^{r-1}(h_j + h_i)} \\
&= \sum_{j=1}^{r-1} \frac{o(h_j^{2r-1})}{\prod_{\substack{i=1 \\ i \neq j}}^{r+1}(h_j - h_i) \prod_{i=1}^{r-1}(h_j + h_i)} \\
&\quad + \sum_{j=r}^{r+1} \frac{\frac{h_j^{2r+1}}{(2r+2)!}\varpi_{2r+2}(f; 0, h_j)}{\prod_{\substack{i=1 \\ i \neq j}}^{r+1}(h_j - h_i) \prod_{i=1}^{r-1}(h_j + h_i)}.
\end{aligned}
$$

Hence,

$$
\begin{aligned}
\lim_{h_{r-1}\to 0} \cdots \lim_{h_1 \to 0} I_2 &= \lim_{h_{r-1}\to 0} \cdots \lim_{h_1 \to 0} \frac{h_r^{2r+1}\varpi_{2r+2}(f; 0, h_r)}{(2r+2)! \prod_{\substack{i=1 \\ \to i \neq r}}^{r+1}(h_r - h_i)\prod_{i=1}^{r-1}(h_r + h_i)} \\
&\quad + \lim_{h_{r-1}\to 0} \cdots \lim_{h_1 \to 0} \frac{h_{r+1}^{2r+1}\varpi_{2r+2}(f; 0, h_{r+1})}{(2r+2)! \prod_{i=1}^{r}(h_{r+1} - h_i)\prod_{i=1}^{r-1}(h_{r+1} + h_i)} \\
&= \frac{1}{(2r+2)!}\left[\frac{h_r^{2r+1}\varpi_{2r+2}(f; 0, h_r)}{h_r^{r-1}(h_r - h_{r+1})h_r^{r-1}} + \frac{h_{r+1}^{2r+1}\varpi_{2r+2}(f; 0, h_{r+1})}{h_{r+1}^{r-1}(h_{r+1} - h_r)h_{r+1}^{r-1}}\right] \\
&= \frac{1}{(2r+2)!}\left[\frac{h_r^{3}\varpi_{2r+2}(f; 0, h_r)}{(h_r - h_{r+1})} + \frac{h_{r+1}^{3}\varpi_{2r+2}(f; 0, h_{r+1})}{(h_{r+1} - h_r)}\right]. \tag{2.3.16}
\end{aligned}
$$

Similarly,

$$
\begin{aligned}
I_4 &= \sum_{j=1}^{r-1} \frac{o(h_j^{2r})}{h_j \prod_{i=1}^{r+1}(h_j + h_i) \prod_{\substack{i=1 \\ i \neq j}}^{r-1}(h_j - h_i)} \\
&= \sum_{j=1}^{r-1} \frac{o(h_j^{2r-1})}{\prod_{i=1}^{r+1}(h_j + h_i) \prod_{\substack{i=1 \\ i \neq j}}^{r-1}(h_j - h_i)},
\end{aligned}
$$

and so

$$\lim_{h_{r-1}\to 0} \ldots \lim_{h_1\to 0} I_4 = 0. \tag{2.3.17}$$

From (2.3.13), (2.3.14), (2.3.16) and (2.3.17),

$$\lim_{h_{r-1}\to 0} \ldots \lim_{h_1\to 0} \frac{Q_{2r}^- + Q_{2r}^+}{2} = \tag{2.3.18}$$

$$\frac{1}{(2r)!} f_{(2r)}^{(s)}(0) + \frac{1}{(2r+2)!} \left[\frac{h_r^3 \varpi_{2r+2}(f;0,h_r)}{(h_r - h_{r+1})} + \frac{h_{r+1}^3 \varpi_{2r+2}(f;0,h_{r+1})}{(h_{r+1} - h_r)} \right].$$

Since the result is true for $k = r$,

$$\lim_{h_{r-1}\to 0} \ldots \lim_{h_1\to 0} Q_{2r}(f; -h_r, \ldots, -h_1, 0, h_1, \ldots, h_r) = \frac{1}{(2r)!} \varpi_{2r}(f;0,h_r).$$

$$\tag{2.3.19}$$

Restoring the values of Q_{2r}^- and Q_{2r}^+ from (2.3.9) and (2.3.10) we get from (2.3.8), (2.3.18) and (2.3.19)

$$\lim_{h_{r-1}\to 0} \ldots \lim_{h_1\to 0} Q_{2r+2}(f; -h_{r+1}, \ldots, -h_1, 0, h_1, \ldots, h_{r+1})$$

$$= \frac{1}{h_{r+1}(h_r + h_{r+1})} \left[\frac{1}{(2r)!} f_{(2r)}^{(s)}(0) + \frac{1}{(2r+2)!} \left[\frac{h_r^3 \varpi_{2r+2}(f;0,h_r)}{(h_r - h_{r+1})} \right. \right.$$

$$\left. \left. + \frac{h_{r+1}^3 \varpi_{2r+2}(f;0,h_{r+1})}{(h_{r+1} - h_r)} \right] - \frac{1}{(2r)!} \varpi_{2r}(f;0,h_r) \right]. \tag{2.3.20}$$

Since,

$$h_r \varpi_{2r+2}(f;0,h_r) = (2r+1)(2r+2) \left[\varpi_{2r}(f;0,h_r) - f_{(2r)}^{(s)}(0) \right]$$

and since $\varpi_{2r}(f;0,h_r) \to f_{(2r)}^{(s)}(0)$ as $h_r \to 0$, we have, taking the limit as $h_r \to 0$ in (2.3.20),

$$\lim_{h_r\to 0} \ldots \lim_{h_1\to 0} Q_{2r+2}(f; -h_{r+1}, \ldots, -h_1, 0, h_1, \ldots, h_{r+1}) = \frac{\varpi_{2r+2}(f;0,h_{r+1})}{(2r+2)!},$$

completing the proof of (a) by induction.

To prove (b), note, as in (2.3.3), that

$$Q_3(f; -h_2, -h_1, h_1, h_2) = \frac{1}{2h_2(h_1 + h_2)} \big[Q_1(f; -h_2, -h_1) + Q_1(f; h_1, h_2)$$

$$-2Q_1(f; -h_1, h_1) \big]$$

$$= \frac{1}{2h_2(h_1 + h_2)} \Big[\frac{f(-h_2) - f(-h_1)}{h_1 - h_2} + \frac{f(h_2) - f(h_1)}{h_2 - h_1}$$

$$-2\frac{f(h_1) - f(-h_1)}{2h_1} \Big]$$

$$= \frac{1}{2h_2(h_1 + h_2)} \Big[\frac{f(h_2) - f(-h_2)}{h_2 - h_1} -$$

$$- \frac{f(h_1) - f(-h_1)}{2h_1} \frac{2h_1}{h_2 - h_1} - 2\frac{f(h_1) - f(-h_1)}{2h_1}$$

and so

$$\lim_{h_1 \to 0} Q_3(f; -h_2, -h_1, h_1, h_2) = \frac{1}{h_2^3} \Big[\frac{f(h_2) - f(-h_2)}{2} - h_2 f_{(1)}^{(s)}(0) \Big]$$

$$= \frac{1}{3!} \varpi_3(f; 0, h_2),$$

proving (b) for $k = 2$. Suppose that it is true for $k = r \geq 2$ and let $k = r + 1$. Then, as in (2.3.8),

$$Q_{2r+1}(f; -h_{r+1}, \ldots, -h_1, h_1, \ldots, h_{r+1})$$

$$= \frac{1}{2h_{r+1}(h_r + h_{r+1})} \big[Q_{2r-1}^- + Q_{2r-1}^+ - 2Q_{2r-1} \big].$$

Then

$$Q_{2r-1}^- = Q_{2r-1}(f; -h_{r+1}, \ldots, -h_1, h_1, \ldots, h_{r-1})$$

$$= \sum_{j=1}^{r+1} \frac{-f(-h_j)}{\prod_{\substack{i=1 \\ i \neq j}}^{r+1}(h_j - h_i) \prod_{i=1}^{r-1}(h_j + h_i)}$$

$$+ \sum_{j=1}^{r-1} \frac{f(h_j)}{\prod_{\substack{i=1 \\ i \neq j}}^{r-1}(h_j - h_i) \prod_{i=1}^{r+1}(h_j + h_i)},$$

and

$$Q_{2r-1}^+ = Q_{2r-1}(f; -h_{r-1}, \ldots, -h_1, h_1, \ldots, h_{r+1})$$

$$= \sum_{j=1}^{r+1} \frac{f(h_j)}{\prod_{\substack{i=1 \\ i \neq j}}^{r+1}(h_j - h_i) \prod_{i=1}^{r-1}(h_j + h_i)}$$

$$+ \sum_{j=1}^{r-1} \frac{-f(-h_j)}{\prod_{\substack{i=1 \\ i \neq j}}^{r-1}(h_j - h_i) \prod_{i=1}^{r+1}(h_j + h_i)},$$

and so

$$\frac{Q_{2r-1}^- + Q_{2r-1}^+}{2}$$

$$= \sum_{j=1}^{r+1} \frac{\frac{1}{2}\left[f(h_j) - f(-h_j)\right]}{\prod_{\substack{i=1 \\ i \neq j}}^{r+1}(h_j - h_i)\prod_{i=1}^{r-1}(h_j + h_i)}$$

$$+ \sum_{j=1}^{r-1} \frac{\frac{1}{2}\left[f(h_j) - f(-h_j)\right]}{\prod_{\substack{i=1 \\ i \neq j}}^{r-1}(h_j - h_i)\prod_{i=1}^{r+1}(h_j + h_i)}.$$

Using the relation

$$\frac{1}{2}\left[f(h_j) - f(-h_j)\right] = \sum_{i=0}^{r-1} \frac{h^{2i+1}}{(2i+1)!}f_{(2i+1)}^{(s)}(0) + \frac{h^{2r+1}}{(2r+1)!}\varpi_{2r+1}(f;0,h_j)$$

and, proceeding as in (a), the proof of (b) for $k = r + 1$ can be completed. $\square$

Theorem 2.3.2 *Let $k \geq 2$.*
(a) If

$$\lim_{h_{k-1} \to 0} \cdots \lim_{h_1 \to 0} Q_{2k}(f; x_0 - h_k, \ldots, x_0 - h_1, x_0, x_0 + h_1, \ldots, x_0 + h_k) \quad (2.3.21)$$

exists finitely, then $f_{(2k-2)}^{(s)}(x_0)$ exists and (2.3.21) equals $\frac{1}{(2k)!}\varpi_{2k}(f; x_0, h_k)$;
(b) if

$$\lim_{h_{k-1} \to 0} \cdots \lim_{h_1 \to 0} Q_{2k-1}(f; x_0 - h_k, \ldots, x_0 - h_1, x_0 + h_1, \ldots, x_0 + h_k) \quad (2.3.22)$$

exists finitely, then $f_{(2k-3)}^{(s)}(x_0)$ exists and (2.3.22) equals $\frac{1}{(2k-1)!}\varpi_{2k-1}(f; x_0, h_k)$.

$\square$ We prove (a). As in Theorem 2.3.1, we may suppose that $x_0 = 0 = f(x_0)$. Since $f(0) = 0$, we have by (1.1.3) of Chapter I

$$Q_{2k}(f; -h_k, \ldots, -h_1, 0, h_1, \ldots, h_k)$$

$$= \sum_{j=1}^{k} \frac{f(-h_j)}{h_j \prod_{\substack{i=1 \\ i \neq j}}^{k}(h_j - h_i)\prod_{i=1}^{k}(h_j + h_i)}$$

$$+ \sum_{j=1}^{k} \frac{f(h_j)}{h_j \prod_{\substack{i=1 \\ i \neq j}}^{k}(h_j - h_i)\prod_{i=1}^{k}(h_j + h_i)}$$

$$= \frac{f(h_1) + f(-h_1)}{2h_1^2 \prod_{i=2}^{k}(h_1^2 - h_i^2)} + \sum_{j=2}^{k} \frac{f(h_j) + f(-h_j)}{2h_j^2 \prod_{\substack{i=1 \\ i \neq j}}^{k}(h_j^2 - h_i^2)}. \quad (2.3.23)$$

Now, by hypothesis, the left-hand side of (2.3.23) tends to a finite limit as $h_1 \to 0$. Also, the summation on the right-hand side of (2.3.23) tends to a finite limit as $h_1 \to 0$. Hence, the first term on the right-hand side of (2.3.23) tends to a finite limit as $h_1 \to 0$. So $\dfrac{f(h_1) + f(-h_1)}{h_1^2}$ tends to a finite limit as $h_1 \to 0$ and this shows, since $f(0) = 0$, that $f_{(2)}^{(s)}(0)$ exists finitely. In addition, if $k = 2$, then from (2.3.23)

$$\lim_{h_1 \to 0} Q_4(f; -h_2, -h_1, 0, h_1, h_2) = -\frac{1}{2h_2^2} f_{(2)}^{(s)}(0) + \frac{f(h_2) + f(-h_2)}{2h_2^4}$$

$$= \frac{1}{h_2^4} \left[\frac{f(h_2) + f(-h_2)}{2} - \frac{h_2^2}{2} f_{(2)}^{(s)}(0) \right] = \frac{1}{4!} \varpi_4(f; 0, h_2),$$

completing the proof in the case $k = 2$.

So, let $k > 2$ and suppose that $f_{(2r)}^{(s)}(0)$ exists finitely, $1 \le r < k - 1$. Then

$$\frac{f(h) + f(-h)}{2} = \sum_{i=1}^{r} \frac{h^{2i}}{(2i)!} f_{(2i)}^{(s)}(0) + \frac{h^{2r+2}}{(2r+2)!} \varpi_{2r+2}(f; 0, h). \qquad (2.3.24)$$

From (2.3.23) and (2.3.24), we have

$$Q_{2k} = Q_{2k}(f; -h_k, \dots, -h_1, 0, h_1, \dots, h_k) = \sum_{j=1}^{k} \frac{f(h_j) + f(-h_j)}{2h_j^2 \prod_{\substack{i=1 \\ i \ne j}}^{k} (h_j^2 - h_i^2)}$$

$$= \sum_{j=1}^{k} \frac{\sum_{i=1}^{r} \frac{h^{2i}}{(2i)!} f_{(2i)}^{(s)}(0)}{h_j^2 \prod_{\substack{i=1 \\ i \ne j}}^{k} (h_j^2 - h_i^2)} + \frac{1}{(2r+2)!} \sum_{j=1}^{k} \frac{h_j^{2r+2} \varpi_{2r+2}(f; 0, h_j)}{h_j^2 \prod_{\substack{i=1 \\ i \ne j}}^{k} (h_j^2 - h_i^2)}.$$

$$(2.3.25)$$

Let $P(x) = \sum_{j=1}^{r} (x^{2i}/(2i)!) f_{(2i)}^{(s)}(0)$. Then $P(0) = 0$ and $P(x) = P(-x)$ so that the first summation on the right-hand side of (2.3.25) is half the divided difference of $P(x)$ at the points $-h_k, \dots, -h_1, 0, h_1, \dots, h_k$. Since $P(x)$ is a polynomial of degree at most $2r < 2k$, this divided difference is zero. So, from (2.3.25),

$$Q_{2k} = \frac{1}{(2r+2)!} \sum_{j=1}^{k} \frac{h_j^{2r} \varpi_{2r+2}(f; 0, h_j)}{\prod_{\substack{i=1 \\ i \ne j}}^{k} (h_j^2 - h_i^2)}. \qquad (2.3.26)$$

Since

$$\frac{h^{2r+2}}{(2r+2)!} \varpi_{2r+2}(f; 0, h) = \frac{f(h) + f(-h)}{2} - \sum_{i=1}^{r} \frac{h^{2i}}{(2i)!} f_{(2i)}^{(s)}(0),$$

we have

$$h^2 \varpi_{2r+2}(f; 0, h) = (2r+2)(2r+1) \frac{(2r)!}{h^{2r}} \left[\frac{f(h) + f(-h)}{2} - \sum_{i=1}^{r} \frac{h^{2i}}{(2i)!} f_{(2i)}^{(s)}(0) \right]$$

$$= (2r+2)(2r+1) \left[\varpi_{2r}(f; 0, h) - f_{(2i)}^{(s)}(0) \right],$$

and so $h^2 \varpi_{2r+2}(f; 0, h) \to 0$ as $h \to 0$. Hence, from (2.3.26),

$$\lim_{h_1 \to 0} Q_{2k} = \frac{1}{(2r+2)!} \sum_{j=2}^{k} \frac{h_j^{2r} \varpi_{2r+2}(f; 0, h_j)}{h_j^2 \prod_{\substack{i=2 \\ i \neq j}}^{k} (h_j^2 - h_i^2)}$$

$$= \frac{1}{(2r+2)!} \sum_{j=2}^{k} \frac{h_j^{2r-2} \varpi_{2r+2}(f; 0, h_j)}{\prod_{\substack{i=2 \\ i \neq j}}^{k} (h_j^2 - h_i^2)},$$

and so

$$\lim_{h_2 \to 0} \lim_{h_1 \to 0} Q_{2k}$$

$$= \frac{1}{(2r+2)!} \sum_{j=3}^{k} \frac{h_j^{2r-2} \varpi_{2r+2}(f; 0, h_j)}{h_j^2 \prod_{\substack{i=3 \\ i \neq j}}^{k} (h_j^2 - h_i^2)} = \frac{1}{(2r+2)!} \sum_{j=3}^{k} \frac{h_j^{2r-4} \varpi_{2r+2}(f; 0, h_j)}{\prod_{\substack{i=3 \\ i \neq j}}^{k} (h_j^2 - h_i^2)}.$$

Continuing this process, we get

$$\lim_{h_r \to 0} \cdots \lim_{h_1 \to 0} Q_{2k} = \frac{1}{(2r+2)!} \sum_{j=r+1}^{k} \frac{\varpi_{2r+2}(f; 0, h_j)}{\prod_{\substack{i=r+1 \\ i \neq j}}^{k} (h_j^2 - h_i^2)}$$

$$= \frac{1}{(2r+2)!} \left[\frac{\varpi_{2r+2}(f; 0, h_{r+1})}{\prod_{i=r+2}^{k} (h_{r+1}^2 - h_i^2)} + \sum_{j=r+2}^{k} \frac{\varpi_{2r+2}(f; 0, h_j)}{\prod_{\substack{i=r+1 \\ i \neq j}}^{k} (h_j^2 - h_i^2)} \right]. \quad (2.3.27)$$

Since $r < k - 1, r + 1 \leq k - 1$ and, so, by the hypothesis $\lim_{h_{r+1} \to 0} \cdots$ $\lim_{h_1 \to 0} Q_{2k}$ exists finitely. Also, $\lim_{h_{r+1} \to 0} \sum_{j=r+2}^{k} \frac{\varpi_{2r+2}(f; 0, h_j)}{\prod_{\substack{i=r+1 \\ i \neq j}}^{k} (h_j^2 - h_i^2)}$ exists finitely. From (2.3.27), $\lim_{h_{r+1} \to 0} \varpi_{2r+2}(f; 0, h_{r+1})$ exists finitely and so $f_{(2r+2)}^{(s)}(0)$ exists finitely.

Hence, by induction on r, we conclude that $f_{(2k-2)}^{(s)}(0)$ exists finitely, completing the proof of the first part of (a).

To prove the second part of (a), note that we have proved (2.3.27) assuming the existence of $f_{(2r)}^{(s)}(0)$ and, therefore, since $f_{(2k-2)}^{(s)}(0)$ exists finitely, (2.3.27) is true if $r = k - 1$ and hence, from (2.3.27) with $r = k - 1$, we have

$$\lim_{h_{k-1} \to 0} \cdots \lim_{h_1 \to 0} Q_{2k} = \frac{1}{2k!} \varpi_{2k}(f; 0, h_k),$$

completing the proof of (a). □

2.4 Cesàro and Peano Derivatives

2.4.1 Cesàro and Peano derivatives, $C_k Df$ and $f_{(k)}$

Theorem 2.4.1 Let f be $C_{r-1}P$-integrable in $[a,b]$. If $x \in [a,b]$, let

$$F_k(x) = \begin{cases} (C_{r-1}P) \int_a^x f, & \text{for } k = 1, \\ (C_{r-k}P) \int_a^x F_{k-1}, & \text{for } 2 \leq k \leq r \end{cases} \tag{2.4.1}$$

and write $\Phi = F_r$. If f is C_r-continuous at x then:

(a) $\Phi_{(r)}(x)$ exists and equals $f(x)$;

(b) $\Phi_{(k)}(x) = F_{r-k}(x)$ if $0 < k < r$;

(c) $C_r(f; x, x+h) = \gamma_r(\Phi; x, h)$ for $x, x+h \in [a,b]$.

□   Since

$$C_r(f; x, x+h) = \frac{r}{h^r}(C_{r-1}P) \int_x^{x+h} (x+h-t)^{r-1} f(t)\, dt,$$

integrating by parts successively we have by Theorem 1.8.1 of Chapter I and using (2.4.1),

$$
\begin{aligned}
\frac{h^r}{r!} C_r(f; x, x+h) &= \frac{1}{(r-1)!}(C_{r-1}P) \int_x^{x+h} (x+h-t)^{r-1} f(t)\, dt \\
&= \frac{1}{(r-1)!} \left[(x+h-t)^{r-1} F_1(t) \right]_x^{x+h} \\
&\quad + \frac{1}{(r-2)!}(C_{r-2}P) \int_x^{x+h} (x+h-t)^{r-2} F_1(t)\, dt \\
&= -\frac{h^{r-1}}{(r-1)!} F_1(x) + \frac{1}{(r-2)!} \left[(x+h-t)^{r-2} F_2(t) \right]_x^{x+h} \\
&\quad + \frac{1}{(r-3)!}(C_{r-3}P) \int_x^{x+h} (x+h-t)^{r-3} F_2(t)\, dt \\
&= -\frac{h^{r-1}}{(r-1)!} F_1(x) - \frac{h^{r-2}}{(r-2)!} F_2(x) + \cdots \\
&= -\sum_{i=1}^{r-1} \frac{h^i}{i!} F_{r-i}(x) + (C_0 P) \int_x^{x+h} F_{r-1} \\
&= \Phi(x+h) - \Phi(x) - \sum_{i=1}^{r-1} \frac{h^i}{i!} F_{r-i}(x). \tag{2.4.2}
\end{aligned}
$$

Since f is C_r-continuous, $C_r(f; x, x + h) \to f(x)$ as $h \to 0$ and, so,

$$\frac{r!}{h^r}\left[\Phi(x + h) - \Phi(x) - \sum_{i=1}^{r-1}\frac{h^i}{i!}F_{r-i}(x)\right] \to f(x) \quad \text{as } h \to 0.$$

This shows that

$$\Phi(x + h) - \Phi(x) - \sum_{i=1}^{r-1}\frac{h^i}{i!}F_{r-i}(x) - \frac{h^r}{r!}f(x) = o(h^r), \text{ as } h \to 0,$$

proving (a) and (b). Also, by (2.4.2), $C_r(f; x, x + h) = \gamma_r(\Phi; x, h)$, which completes the proof. $\qquad\square$

Theorem 2.4.2 *Let f be C_r-continuous in $[a, b]$ and let Φ be the function of Theorem 2.4.1. Then, for $x \in [a, b]$, $f(x) = \Phi_{(r)}(x)$ and $\overline{C_r D}f(x) = \overline{\Phi}_{(r+1)}(x)$, etc.; further, $C_r Df(x)$ exists if and only if $\Phi_{(r+1)}(x)$ exits and in either case they are equal.*

$\square$ The first part of this theorem follows from Theorem 2.4.1 (c), and using parts (c) and (a) of that theorem,

$$\frac{r+1}{h}\left[C_r(f; x, x + h) - f(x)\right] = \frac{r+1}{h}\left[\gamma_r(\Phi; x, h) - \Phi_{(r)}(x)\right] = \gamma_{r+1}(\Phi; x, h),$$

the result follows. $\qquad\square$

Theorem 2.4.3 *Let Φ be the function of Theorem 2.4.1. If $\Phi_{(r)}$ exists finitely in $[a, b]$, then for $1 \le k \le r$*

(a) $\Phi_{(k)}$ *is the C_{k-1}-derivative of $\Phi_{(k-1)}$ in $[a, b]$;*

(b) $\Phi_{(k)}$ *is C_k-continuous in $[a, b]$;*

(c) $\Phi_{(k-1)}$ *is an indefinite $C_{k-1}P$ integral of $\Phi_{(k)}$ in $[a, b]$;*

(d) $C_k(f; x, x + h) = \gamma_k(\Phi; x, h)$ *for $x, x + h \in [a, b]$.*

$\square$ If $r = 1$, then $\Phi_{(1)}$, being the ordinary first derivative of Φ, is the C_0-derivative of Φ, $\Phi_{(1)}$ is C_0P-integrable and $\Phi_{(0)} = \Phi$ is its indefinite C_0P-integral; further,

$$C_1(\Phi_{(1)}; x, x + h) = \frac{1}{h}(C_0P) - \int_x^{x+h}\Phi_{(1)} = \frac{\Phi(x + h) - \Phi(x)}{h} = \gamma_1(\Phi; x, x + h).$$

Hence, $C_1(\Phi_{(1)}; x, x + h) \to \Phi_{(1)}(x)$ as $h \to 0$, proving the result in the case $r = 1$.

Suppose now that the result is true for $r = 1, 2, \ldots, r_0$, we prove the result for $r = r_0 + 1$. Let $\Phi_{(r_0+1)}$ exist in $[a, b]$. Then $\Phi_{(r_0)}$ exists in $[a, b]$. By the induction hypothesis, $\Phi_{(r_0)}$ is the C_{r_0-1}-derivative of $\Phi_{(r_0-1)}$, $\Phi_{(r_0)}$ is C_{r_0}-continuous, $\Phi_{(r_0-1)}$ is the indefinite $C_{r_0-1}P$-integral of $\Phi_{(r_0)}$ and $C_{r_0}(f; x, x + h) = \gamma_{r_0}(\Phi; x, h)$.

Hence,

$$\frac{r_0 + 1}{h}\left[C_{r_0}(\Phi_{(r_0)}; x, x + h) - \Phi_{(r_0)}(x)\right] = \frac{r_0 + 1}{h}\left[\gamma_{r_0}(\Phi; x, h) - \Phi_{(r_0)}(x)\right]$$
$$= \gamma_{r_0+1}(\Phi; x, h),$$

and, since $\Phi_{(r_0+1)}(x)$ exists, $\gamma_{r_0+1}(\Phi; x, h) \to \Phi_{(r_0+1)}(x)$ as $h \to 0$ and so,

$$\lim_{h \to 0} \frac{r_0 + 1}{h}\left[C_{r_0}(\Phi_{(r_0)}; x, x + h) - \Phi_{(r_0)}(x)\right] = \Phi_{(r_0+1)}(x).$$

Thus, $C_{r_0}D\Phi_{(r_0)}(x) = \Phi_{(r_0+1)}(x)$ and, since this holds for all $x \in [a, b]$, it follows, from the definition of the $C_r P$-integral, that $\Phi_{(r_0+1)}$ is $C_{r_0}P$-integrable and $\Phi_{(r_0)}$ is its indefinite $C_{r_0}P$-integral. So, integrating by parts successively as in (2.4.2),

$$\frac{h^{r_0+1}}{(r_0 + 1)!}C_{r_0+1}(\Phi_{(r_0+1)}; x, x + h) = \tfrac{1}{r_0!}(C_{r_0}P) - \int_x^{x+h}(x + h - t)^{r_0}\Phi_{(r_0+1)}(t)\,dt$$
$$= \Phi(x + h) - \Phi(x) - \sum_{i=1}^{r_0}\tfrac{h^i}{i!}\Phi_{(i)}(x).$$

Hence, $C_{r_0+1}(\Phi_{(r_0+1)}; x, x + h) = \gamma_{r_0+1}(\Phi; x, h)$, thus $C_{r_0+1}(\Phi_{(r_0+1)}; x, x + h) \to \Phi_{(r_0+1)}(x)$, proving that $\Phi_{(r_0+1)}$ is C_{r_0+1}-continuous at x. So, the result is true for $r = r_0 + 1$ and the proof is now completed by induction. $\qquad\square$

Theorem 2.4.4 *Let Φ be the function of Theorem 2.4.1 and let $\Phi_{(r)}$ exist finitely in $[a, b]$ and let $x \in [a, b]$. Then $\Phi_{(r+1)}(x)$ exists if and only if $C_r D\Phi_r(x)$ exists and in either case they are equal.*

$\square$ From (d) of Theorem 2.4.3, we have

$$\frac{r+1}{h}\left[C_r(\Phi_{(r)}; x, x+h) - \Phi_{(r)}(x)\right] = \frac{r+1}{h}\left[\gamma_r(\Phi; x, h) - \Phi_{(r)}(x)\right] = \gamma_{r+1}(\Phi; x, h),$$

and the result follows. $\qquad\qquad\qquad\qquad\qquad\qquad\qquad\qquad\qquad\qquad\square$

Remark. Remark. If $\Phi_{(r)}$ exists at a single point, say ξ, then it may be the case that the previous derivatives $\Phi_{(k)}$, $1 \le k \le r-1$, do not exist at any point except ξ and it may even be the case that the function Φ is not measurable and, so, the existence of the Cesàro derivative at ξ is not guaranteed.

To see this, let $A \subset I = [-\tfrac{1}{2}, \tfrac{1}{2}]$ be a nonmeasurable set such that its outer measure and that of its complement in I are both equal to 1. It follows that there is a point $\xi \in I$ such that every neighbourhood $N(\xi)$ of ξ contains a portion of A and its complement. Define

$$\Phi(x) = \begin{cases} (x - \xi)^{r+1} & \text{if } x \in A, \\ (x - \xi)^{r+2} & \text{if } x \notin A. \end{cases}$$

Then

$$\Phi(\xi + h) = \begin{cases} h^{r+1} & \text{if } \xi + h \in A, \\ h^{r+2} & \text{if } \xi + h \notin A. \end{cases}$$

So, $\Phi(\xi + h) = o(h^r)$ as $h \to 0$, and hence, $\Phi_{(r)}(\xi)$ exists. However, there is no neighbourhood of ξ in which $\Phi_{(k)}$ exists, $1 \le k \le r$, except at ξ. Clearly Φ is not measurable.

2.4.2   Cesàro and absolute Peano derivatives, $C_k Df$ and f^*

Theorem 2.4.5 *Let f be defined in a neighbourhood $N(x_0)$ of x_0 in $\mathbb{R}$. If $f^*(x_0)$ exists, then $C_n Df(x_0)$ exists for some n and $C_n Df(x_0) = f^*(x_0)$. The converse is not true.*

☐   Suppose that $f^*(x_0)$ exists. Then there is a function g and a nonnegative integer n such that

$$g_{(n)}(x) = f(x), \ x \in N(x_0), \tag{2.4.3}$$
$$g_{(n+1)}(x_0) = f^*(x_0). \tag{2.4.4}$$

By Theorem 2.4.4, $g_{(n+1)}(x_0)$ is the Cesàro derivative of $g_{(n)}$ at x_0 of order n. Hence, by (2.4.3) and (2.4.4), $f^*(x_0)$ is the Cesàro derivative $C_n Df(x_0)$. This proves the first part.

For the second part, consider the following example.

Let $E = \{0\} \cup \left\{ \sqrt{\frac{2}{(4n+1)\pi}} ; n = 1, 2, 3, \ldots \right\}$ and define

$$f(x) = \begin{cases} 0, & \text{if } x \in E, \\ \sin(1/x^2), & \text{if } x \in \mathbb{R} \setminus E. \end{cases}$$

Then f is Riemann integrable and set $F(x) = \int_0^x f$. Changing the variable and applying the second mean value theorem,

$$|F(x)| = \left| \int_0^x \sin(1/t^2) \, dt \right| = \frac{1}{2} \left| \int_{x^{-2}}^{\infty} \frac{\sin u}{u^{3/2}} \, du \right| = \frac{|x^3|}{2} \left| \int_{x^{-2}}^{\xi} \sin u \, du \right| \le |x^3|.$$

This implies that $F_{(1)}(0) = F_{(2)}(0) = 0$, and $F_{(1)}(x) = \sin(1/x^2)$, $x \ne 0$. Hence,

$$\lim_{h \to 0} \frac{C_1(f; 0, h) - f(0)}{\frac{1}{2}h} = \lim_{h \to 0} \frac{\frac{1}{h}\int_0^h f - f(0)}{\frac{1}{2}h}$$

$$= \lim_{h \to 0} \frac{2}{h^2} [F(h) - F(0) - F_{(1)}(0)] = F_{(2)}(0).$$

So, $C_1 Df(0)$ exists with value 0. However, $f^*(0)$ does not exist. For, if possible, suppose the contrary, that $f^*(0)$ does exist. Let $\delta > 0$ be arbitrary and choose n such that $\sqrt{\frac{1}{2n\pi}} < \delta$. Then putting $x_1 = \sqrt{\frac{2}{(4n+1)\pi}}$, $x_2 = \sqrt{\frac{4}{(8n+1)\pi}}$, we have from the definition of f that $f(x_1) = 0$, $f(x_2) = \frac{1}{\sqrt{2}}$ and $f(x) > \frac{1}{\sqrt{2}}$ for $x_1 < x < x_2$. Hence, f does not satisfy the Darboux property on the interval $I = (-\delta, \delta)$. Therefore, there is no function g such that for some n, $g_{(n)}(x) = f(x)$ in I since a Peano derivative has the Darboux property. Since δ was arbitrary, $f^*(0)$ cannot exist.

Remark. Remark Since every Peano derivative is a generalized Peano derivative, in view of the Remark in 3.1, if f has generalized Peano derivative at x_0, the Cesàro derivative of f at x_0 may not exist.

2.5   Peano and Symmetric de la Vallée Poussin Derivatives, $f_{(k)}$ and $f_{(k)}^{(s)}$, and Smoothness of Order k

Theorem 2.5.1 *If $f_{(k)}(x)$ exists finitely, then $f_{(k)}^{(s)}(x)$ exists finitely and the two derivatives are equal. Moreover,*

(a) $\dfrac{\left[\gamma_{k+1}(f;x,h) + \gamma_{k+1}(f;x,-h)\right]}{2} = \varpi_{k+1}(f;x,h);$

(b) $\dfrac{\left[\gamma_{k+1}(f;x,h) - \gamma_{k+1}(f;x,-h)\right]}{2} = \dfrac{h}{k+2}\varpi_{k+2}(f;x,h).$

☐   Let $f_{(k)}(x)$ exist finitely and suppose that k is even, $k = 2m$, say. Then, since

$$f(x+h) = \sum_{i=0}^{k} \frac{h^i}{i!} f_{(i)}(x) + o(h^k) \text{ as } h \to 0,$$

we have

$$\frac{f(x+h) + f(x-h)}{2} = \frac{1}{2}\sum_{i=0}^{2m} \frac{h^i + (-h)^i}{i!} f_{(i)}(x) + o(h^{2m})$$

$$= \sum_{j=0}^{m} \frac{h^{2j}}{(2j)!} f_{(2j)}(x) + o(h^{2m}),$$

which shows that $f_{(2m)}^{(s)}(x)$ exists and $f_{(2j)}^{(s)}(x) = f_{(2j)}(x)$, $j = 1, 2, \ldots, m$. Also, since

$$\gamma_{k+1}(f;x,h) = \frac{(k+1)!}{h^{k+1}}\left[f(x+h) - \sum_{i=0}^{k} \frac{h^i}{i!} f_{(i)}(x)\right],$$

we have that

$$
\frac{\left[\gamma_{k+1}(f;x,h) + \gamma_{k+1}(f;x,-h)\right]}{2} = \frac{(2m+1)!}{h^{2m+1}} \left[\frac{f(x+h) - f(x-h)}{2} \right.
$$

$$
\left. -\frac{1}{2} \sum_{i=0}^{2m} \frac{h^i - (-h)^i}{i!} f_{(i)}(x) \right]
$$

$$
= \frac{(2m+1)!}{h^{2m+1}} \left[\frac{f(x+h) - f(x-h)}{2} - \sum_{j=0}^{m-1} \frac{h^{2j+1}}{(2j+1)!} f_{(2j+1)}(x) \right]
$$

$$
= \varpi_{2m+1}(f;x,h)
$$

and that

$$
\frac{\left[\gamma_{k+1}(f;x,h) - \gamma_{k+1}(f;x,-h)\right]}{2} = \frac{(2m+1)!}{h^{2m+1}} \left[\frac{f(x+h) + f(x-h)}{2} \right.
$$

$$
\left. -\frac{1}{2} \sum_{i=0}^{2m} \frac{h^i + (-h)^i}{i!} f_{(i)}(x) \right]
$$

$$
= \frac{(2m+1)!}{h^{2m+1}} \left[\frac{f(x+h) + f(x-h)}{2} - \sum_{j=0}^{m} \frac{h^{2j}}{(2j)!} f_{(2j)}(x) \right]
$$

$$
= \frac{h}{2m+2} \varpi_{2m+2}(f;x,h).
$$

This proves the result for k even.

Let now k be odd, $k = 2m + 1$, say. Then, as above,

$$
\frac{f(x+h) - f(x-h)}{2} = \frac{1}{2} \sum_{i=0}^{2m+1} \frac{h^i - (-h)^i}{i!} f_{(i)}(x) + o(h^{2m+1})
$$

$$
= \sum_{j=0}^{m} \frac{h^{2j+1}}{(2j+1)!} f_{(2j+1)}(x) + o(h^{2m+1}),
$$

showing that $f_{(2m+1)}^{(s)}(x)$ exists and $f_{(2j+1)}^{(s)}(x) = f_{(2j+1)}(x)$, $j = 0, 1, \ldots, m$. Also,

$$
\frac{\left[\gamma_{k+1}(f;x,h) + \gamma_{k+1}(f;x,-h)\right]}{2} = \frac{(2m+2)!}{h^{2m+2}} \left[\frac{f(x+h) + f(x-h)}{2} \right.
$$

$$
\left. -\frac{1}{2} \sum_{i=0}^{2m+1} \frac{h^i + (-h)^i}{i!} f_{(i)}(x) \right]
$$

$$
= \frac{(2m+2)!}{h^{2m+2}} \left[\frac{f(x+h) + f(x-h)}{2} - \sum_{j=0}^{m} \frac{h^{2j}}{(2j)!} f_{(2j)}(x) \right]
$$

$$
= \varpi_{2m+2}(f;x,h)
$$

and

$$\frac{\left[\gamma_{k+1}(f;x,h) - \gamma_{k+1}(f;x,-h)\right]}{2} = \frac{(2m+2)!}{h^{2m+2}}\left[\frac{f(x+h) - f(x-h)}{2}\right.$$

$$\left. -\frac{1}{2}\sum_{i=0}^{2m+1}\frac{h^i - (-h)^i}{i!}f_{(i)}(x)\right]$$

$$= \frac{(2m+2)!}{h^{2m+2}}\left[\frac{f(x+h) - f(x-h)}{2} - \sum_{j=0}^{m}\frac{h^{2j+1}}{(2j+1)!}f_{(2j+1)}(x)\right]$$

$$= \frac{h}{2m+3}\varpi_{2m+3}(f;x,h),$$

showing that the result is also true when k is odd. □

Corollary 2.5.2 *If $f_{(k)}(x)$ exists, possibly infinitely, then $f_{(k)}^{(s)}(x)$ exists with the same value.*

□   The proof follows from Theorem 2.5.1(a).   □

The converse of Theorem 2.5.1 is not true. Suppose that k is even and

$$f(x) = \begin{cases} \sin 1/x, & \text{if } x \neq 0, \\ 0, & \text{if } x = 0. \end{cases}$$

Then $f_{(k)}^{(s)}(0) = 0$, but $f_{(k)}(0)$ does not exist since $f_{(1)}(0)$ does not exist. If k is odd, replace the sine by cosine in the above example to get a similar result.

However, we have the following result.

Theorem 2.5.3 *If f is smooth at x of orders $k+1$ and $k+2$, then $f_{(k)}(x)$ exists finitely. If $f_{(k)}(x)$ exists finitely, then f is smooth at x of order $k+1$. More generally, if the Peano derivates $\overline{f}_{(k+1)}^{+}(x), \overline{f}_{(k+1)}^{-}(x), \underline{f}_{(k+1)}^{+}(x), \underline{f}_{(k+1)}^{-}(x)$ are all finite, then*

$$\limsup_{h\to 0+} h\varpi_{k+2}(f;x,h) \leq (k+2)\frac{\overline{f}_{(k+1)}^{+}(x) - \underline{f}_{(k+1)}^{-}(x)}{2},$$

$$\liminf_{h\to 0+} h\varpi_{k+2}(f;x,h) \geq (k+2)\frac{\underline{f}_{(k+1)}^{+}(x) - \overline{f}_{(k+1)}^{-}(x)}{2}.$$

□   Since f is smooth of order $k+1$ at x, $f_{(k-1)}^{(s)}(x)$ exists finitely and $h\varpi_{k+1}(f;x,h) = o(1)$ as $h \to 0$. That is, if k is even, $k = 2m$, say,

$$\frac{(2m+1)!}{h^{2m}}\left[\frac{f(x+h) - f(x-h)}{2} - \sum_{i=0}^{m-1}\frac{h^{2i+1}}{(2i+1)!}f_{(2i+1)}^{(s)}(x)\right] = o(1) \text{ as } h \to 0,$$

which gives

$$\frac{f(x+h) - f(x-h)}{2} = \sum_{i=0}^{m-1} \frac{h^{2i+1}}{(2i+1)!} f_{(2i+1)}^{(s)}(x) + o(h^{2m}) \text{ as } h \to 0. \quad (2.5.1)$$

Similarly, smoothness of order $k + 2 = 2m + 2$ gives

$$\frac{f(x+h) + f(x-h)}{2} = \sum_{i=0}^{m} \frac{h^{2i}}{(2i)!} f_{(2i)}^{(s)}(x) + o(h^{2m+1}) \text{ as } h \to 0. \quad (2.5.2)$$

From (2.5.1) and (2.5.2),

$$f(x+h) = \sum_{i=0}^{m} \frac{h^{2i}}{(2i)!} f_{(2i)}^{(s)}(x) + \sum_{i=0}^{m-1} \frac{h^{2i+1}}{(2i+1)!} f_{(2i+1)}^{(s)}(x) + o(h^{2m}) \text{ as } h \to 0,$$

which shows that $f_{(2m)}(x)$ exists. If k is odd, the argument is similar.

Now suppose that $f_{(k)}(x)$ exists finitely. Then

$$h\gamma_{k+1}(f; x, h) = (k+1)\big[\gamma_k(f; x, h) - f_{(k)}(x)\big] \quad (2.5.3)$$

and since the right-hand side of (2.5.3) tends to 0 as $h \to 0$, $h\gamma_{k+1}(f; x, h) = o(1)$; and so, from Theorem 2.5.1(a), $h\varpi_{k+1}(f; x, h) = o(1)$, showing that f is smooth of order $k + 1$ at x. The last two inequalities follow from Theorem 2.5.1(b). □

Theorem 2.5.4 *If $f_{(k)}(x)$ exists finitely and if f is smooth of order $k + 2$ at x, then*

$$\overline{f}_{(k)}^+(x) = \overline{f}_{(k)}^-(x) \text{ and } \underline{f}_{(k)}^+(x) = \underline{f}_{(k)}^-(x).$$

□   From Theorem 2.5.1(b),

$$\gamma_{k+1}(f; x, h) = \frac{2h}{k+2}\varpi_{k+2}(f; x, h) + \gamma_{k+1}(f; x, -h),$$

and so the result follows by letting $h \to 0+$. □

Theorem 2.5.5 *If f is Peano bounded of order r at x, then f is d.l.V.P. bounded of order r at x.*

□   By Theorem 1.4.4 of Chapter I, $f_{(r-1)}(x)$ exists finitely and so, by Theorem 2.5.1(a), the result follows. □

2.6 Symmetric Cesàro and Symmetric de la Vallée Poussin Derivatives, SC_kDf and $f_{(k)}^{(s)}$

Theorem 2.6.1 *Let f be $C_{r-1}P - integrable$ in $[a, b]$ and for $x \in [a, b]$ let*

$$F_k(x) = \begin{cases} (C_{r-1}P) \int_a^x f, & \text{if } k = 1, \\ (C_{r-k}P) \int_a^x F_{k-1}, & \text{if } 2 \le k \le r, \end{cases}$$

and write $\Phi = F_r$. Then the Peano derivative $\Phi_{(r-1)}$ exists in $[a, b]$ and for $x \in [a, b]$

$$\frac{r+1}{2h}\Big[C_r(f; x, x+h) - C_r(f; x, x-h)\Big] = \varpi_{r+1}(\Phi; x, h). \qquad (2.6.1)$$

☐   As in (2.4.2), we have

$$\frac{h^r}{r!}C_r(f; x, x+h) = \Phi(x+h) - \Phi(x) - \sum_{i=1}^{r-1}\frac{h^i}{i!}F_{r-i}(x),$$

and so

$$C_r(f; x, x+h) - C_r(f; x, x-h) =$$
$$\frac{r!}{h^r}\Big[\Phi(x+h) - \Phi(x) - \sum_{i=1}^{r-1}\frac{h^i}{i!}F_{r-i}(x)\Big]$$
$$-\frac{r!}{(-h)^r}\Big[\Phi(x-h) - \Phi(x) - \sum_{i=1}^{r-1}\frac{(-h)^i}{i!}F_{r-i}(x)\Big]$$
$$= \frac{r!}{h^r}\Big[\Phi(x+h) - (-1)^r\Phi(x-h) - \Phi(x) + (-1)^r\Phi(x)$$
$$-\sum_{i=1}^{r-1}\frac{h^i}{i!}\big(1 - (-1)^{r-i}\big)F_{r-i}(x)\Big].$$

Hence,

$$\frac{r+1}{2h}\big[C_r(f; x, x+h) - C_r(f; x, x-h)\big] =$$
$$\frac{(r+1)!}{h^{r+1}}\Big[\frac{\Phi(x+h) - (-1)^r\Phi(x-h)}{2}\frac{1 - (-1)^r}{2}\Phi(x)$$
$$-\sum_{i=1}^{r-1}\frac{1 - (-1)^{r-i}}{2}\frac{h^i}{i!}F_{r-i}(x)\Big].$$

So, if r is even, $r = 2m$ say, then

$$\frac{r+1}{2h}\left[C_r(f;x,x+h) - C_r(f;x,x-h)\right] =$$

$$\frac{(2m+1)!}{h^{2m+1}}\left[\frac{\Phi(x+h) - \Phi(x-h)}{2} - \sum_{i=1}^{2m-1}\frac{1-(-1)^{2m-i}}{2}\frac{h^i}{i!}F_{2m-i}(x)\right]$$

$$= \frac{(2m+1)!}{h^{2m+1}}\left[\frac{\Phi(x+h) - \Phi(x-h)}{2} - \sum_{j=0}^{m-1}\frac{h^{2j+1}}{(2j+1)!}F_{2m-2j-1}(x)\right]. \quad (2.6.3)$$

Since a $C_{r-k}P$-integral is $C_{r-k-1}P$-integrable and C_{r-k}-continuous, and since F_{r-i} is a C_iP-integral, F_{r-i} is $C_{i-1}P$-integrable and C_i-continuous, and so, in particular, F_1 is $C_{r-2}P$-integrable and C_{r-1}-continuous in $[a,b]$. Hence, by Theorem 2.4.1, $\Phi_{(r-1)}$ exists and $\Phi_{(k)} = F_{r-k}$ in $[a,b]$ for $1 \le k \le r-1$. Therefore, $\Phi_{(2j+1)} = F_{2m-2j-1}$ for $0 \le j \le m-1$, and so from (2.6.3)

$$\frac{r+1}{2h}\left[C_r(f;x,x+h) - C_r(f;x,x-h)\right] =$$

$$\frac{(2m+1)!}{h^{2m+1}}\left[\frac{\Phi(x+h) - \Phi(x-h)}{2} - \sum_{j=0}^{m-1}\frac{h^{2j+1}}{(2j+1)!}\Phi_{(2j+1)}(x)\right]. \quad (2.6.4)$$

Similarly, if r is odd, $r = 2m+1$ say, then from (2.6.2)

$$\frac{r+1}{2h}\left[C_r(f;x,x+h) - C_r(f;x,x-h)\right] =$$

$$\frac{(2m+2)!}{h^{2m+2}}\left[\frac{\Phi(x+h) + \Phi(x-h)}{2} - \sum_{i=0}^{2m}\frac{1-(-1)^{2m-i+1}}{2}\frac{h^i}{i!}F_{2m-i+1}(x)\right],$$

$$\quad (2.6.5)$$

and so, as above, $\Phi_{(2j)} = F_{2m-2j+1}$ and hence, from (2.6.5)

$$\frac{r+1}{2h}\left[C_r(f;x,x+h) - C_r(f;x,x-h)\right] =$$

$$\frac{(2m+2)!}{h^{2m+2}}\left[\frac{\Phi(x+h) + \Phi(x-h)}{2} - \sum_{j=0}^{m}\frac{h^{2j}}{(2j)!}\Phi_{(2j)}(x)\right]. \quad (2.6.6)$$

Hence, from (2.6.4) and (2.6.6) and from the definition of $\varpi_{r+1}(\Phi;x,h)$, we obtain (2.6.1). $\quad\Box$

Corollary 2.6.2 *Let the hypotheses of the above theorem hold. Then, f is SC_r-continuous at x if and only if Φ is smooth of order $r+1$ at x. Also, $\overline{SCD}_r f(x) = \overline{\Phi}_{(r+1)}^{(s)}(x), \underline{SCD}_r f(x) = \underline{\Phi}_{(r+1)}^{(s)}(x)$ and $SC_r Df(x)$ exists if and only if $\Phi_{(r+1)}^{(s)}(x)$ exists.*

$\Box$ The proof follows from (2.6.1). $\quad\Box$

2.7　Borel and Peano Derivatives, $BD_k f$ and $f_{(k)}$

Theorem 2.7.1 *Let f be special Denjoy integrable in some neighbourhood of x. If $f_{(r)}(x)$ exists, then $BD_r f(x)$ exists and the two derivatives are equal, but the converse is not true. More generally, if $f_{(r-1)}(x)$ exists finitely, then*

$$\underline{f}_{(r)}(x) \le \underline{BD}_r f(x) \le \overline{BD}_r f(x) \le \overline{f}_{(r)}(x).$$

□　　If $f_{(r)}(x)$ exists finitely, then

$$f(x + t) = \sum_{i=0}^{r} \frac{t^i}{i!} f_{(i)}(x) + o(t^r) \text{ as } t \to 0,$$

and so

$$\frac{1}{h} \int_0^h \frac{\left[f(x+t) - \sum_{i=0}^{r} \frac{t^i}{i!} f_{(i)}(x) \right]}{t^r} \, dt = o(1) \text{ as } h \to 0.$$

Hence, $BD_r f(x)$ exists and $BD_r f(x) = f_{(r)}(x)$.

If $f_{(r)}(x)$ is infinite, then $f_{(i)}(x)$ exists finitely, $0 \le i \le r - 1$, and so, by the above argument, $BD_i f(x)$ exists and is equal to $f_{(i)}(x)$, $0 \le i \le r - 1$.

Suppose that $f_{(r)}(x) = \infty$. Then for any $M > 0$ there is a $\delta > 0$ such that

$$\frac{\left[f(x+t) - \sum_{i=0}^{r-1} \frac{t^i}{i!} f_{(i)}(x) \right]}{t^r / r!} > M \text{ for } 0 < |t| < \delta.$$

Since $f_{(i)}(x) = BD_i f(x)$, $0 \le i \le r - 1$, this gives

$$\frac{r!}{h} \int_\epsilon^h \frac{\left[f(x+t) - \sum_{i=0}^{r-1} \frac{t^i}{i!} BD_i f(x) \right]}{t^r} \, dt > M \frac{h - \epsilon}{h} \text{ for } 0 < \epsilon < h < \delta.$$

Letting $\epsilon \to 0+$ first and then $h \to 0+$, we have that $\underline{BD}_r^+ f(x) \ge M$.

Similarly, $\underline{BD}_r^- f(x) \ge M$ and so $\underline{BD}_r f(x) \ge M$.

M being arbitrary this shows that $\underline{BD}_r f(x) = \infty$. Hence, $BD_r f(x)$ exists and equals $f_{(r)}(x)$.

If $f_{(r)}(x) = -\infty$, the proof is similar.

Regarding the converse, let r be a fixed positive integer and consider

$$f(x) = \begin{cases} \sin x, & \text{if } x \text{ is rational,} \\ x^r \sin(1/x^{r-1}), & \text{if } x \text{ is irrational.} \end{cases}$$

Then $f'(0)$ does not exist and so $f_{(r)}(0)$ cannot exist. To see that $BD_r(0)$ exists first note that

$$\frac{1}{h} \int_0^h \frac{f(t)}{t^r} \, dt = \frac{1}{h} \int_0^h \sin(1/t^{r-1}) \, dt. \tag{2.7.1}$$

Now if $r \geq 2$,

$$\left(t^r \cos(1/t^{r-1})\right)' = r t^{r-1} \cos(1/t^{r-1}) + (r-1)\sin(1/t^{r-1}). \qquad (2.7.2)$$

The right-hand side of (2.7.2) being Riemann integrable,

$$(r-1)\int_0^h \sin(1/t^{r-1})\,dt = h^r \cos(1/h^{r-1}) - r\int_0^h t^{r-1}\cos(1/t^{r-1})\,dt$$

and hence,

$$\frac{1}{h}\int_0^h \sin(1/t^{r-1})\,dt = \frac{h^{r-1}}{r-1}\cos(1/h^{r-1}) - \frac{r}{r-1}\frac{1}{h}\int_0^h t^{r-1}\cos(1/t^{r-1})\,dt.$$
$$(2.7.3)$$

The right-hand side of (2.7.3) tends to 0 as $h \to 0$ and so

$$\frac{1}{h}\int_0^h \sin(1/t^{r-1})\,dt = o(1) \text{ as } h \to 0. \qquad (2.7.4)$$

From (2.7.1) and (2.7.4) and the definition of $BD_r f$, it follows that $BD_i(0)$ exists $0 \leq i \leq r$.

To prove the last part, we consider the last inequality, the consideration of the first being similar.

We may then assume that $\overline{f}_{(r)}(x) < \infty$, $\overline{f}_{(r)}(x) < M$, say. Then there is a $\delta > 0$ such that $\gamma_r(f;x,t) < M$ for $0 < |t| < \delta$, where $\gamma_r(f;x,t)$ is as defined in (1.4.3) of Chapter I. Hence,

$$\frac{1}{h}\int_\epsilon^h \gamma_r(f;x,t)\,dt \leq M\frac{h-\epsilon}{h} \text{ for } 0 < \epsilon < h < \delta. \qquad (2.7.5)$$

As $f_{(r-1)}(x)$ exists finitely, we have by the first part that $f_{(r-1)}(x) = BD_{r-1}f(x)$, and a fortiorti $f_{(i)}(x) = BD_i f(x)$, $0 \leq i \leq r-1$. So from (2.7.5) by letting $\epsilon \to 0+$ and then $h \to 0+$, we have that $\overline{BD}_r^+ f(x) \leq M$, Similarly, $\overline{BD}_r^- f(x) \leq M$ showing that $\overline{BD}_r f(x) \leq M$. Since M is arbitrary, the inequality is proved. $\qquad \square$

Theorem 2.7.2 *Let f be special Denjoy integrable in some neighbourhood of x with F its indefinite integral in that neighbourhood. If $F_{(r+1)}(x)$ exists, then $BD_r f(x)$ exists with $BD_r f(x) = F_{(r+1)}(x)$ and*

$$\underline{F}_{(r+2)}(x) \leq \underline{BD}_{r+1}f(x) \leq \overline{BD}_{r+1}f(x) \leq \overline{F}_{(r+2)}(x). \qquad (2.7.6)$$

To prove this theorem, we need the following:

Lemma 2.7.3 *If $\lim_{t\to 0}\phi(x+t) = \ell$, then $\lim_{h\to 0}\frac{1}{h}\int_0^h \phi(x+t)\,dt = \ell$, where $-\infty \leq \ell \leq \infty$. More generally, if $\limsup_{h\to 0}\phi(x+t) = M$, $-\infty < M \leq \infty$, then $\limsup_{h\to 0}\frac{1}{h}\int_0^h \phi(x+t)\,dt \leq M$. There is a similar result for $\liminf$.*

□   We first prove the second part of the lemma. We may without loss in generality suppose that $M < \infty$. If then $M' > M$, there is a $\delta > 0$ such that $\phi(x+t) < M'$ if $|t| < \delta$. Hence, if $0 < |h| < \delta$, then $\frac{1}{h} \int_0^h \phi(x+t)\,dt < M'$ and, hence, $\limsup_{h \to 0} \frac{1}{h} \int_0^h \phi(x+t)\,dt \leq M'$. Since M' is arbitrary, this proves this part of the lemma.

The proof of the lim inf case is similar.

Now if ℓ is finite, the proof of the the first statement is similar to the above proofs. Suppose then $\ell = \infty$ and let $G > 0$. Then there is a $\delta > 0$ such that $\phi(x+t) > G$ if $|t| < \delta$. So, if $0 < |h| < \delta$, then $\frac{1}{h} \int_0^h \phi(x+t)\,dt > G$ and, since G was arbitrary, this proves that $\lim_{t \to 0} \phi(x+t)\,dt = \infty$. The case when $\ell = -\infty$ is similar.   □

Proof of Theorem 2.7.2:

□   Let $r = 1$.

Since $F_{(2)}(x)$ exists, $F_{(1)}(x)$ exists finitely and

$$F_{(1)}(x) = \lim_{h \to 0} \frac{F(x+h) - F(x)}{h} = \lim_{h \to 0} \frac{1}{h} \int_0^h f(x+t)\,dt = BD_0 f(x)$$

by definition. Let $0 < \epsilon < h$. Then integrating by parts,

$$\int_\epsilon^h \frac{f(x+t) - BD_0 f(x)}{t}\,dt \tag{2.7.7}$$

$$= \frac{F(x+t) - F(x) - tBD_0(x)}{t}\bigg|_\epsilon^h + \int_\epsilon^h \frac{F(x+t) - F(x) - tBD_0 f(x)}{t^2}\,dt.$$

If $F_{(2)}(x)$ is finite, the right-hand side of (2.7.7) tends to a limit as $\epsilon \to 0$. So, letting $\epsilon \to 0$ and dividing by h, we have from (2.7.7)

$$\frac{1}{h} \int_0^h \frac{f(x+t) - BD_0 f(x)}{t}\,dt \tag{2.7.8}$$

$$= \frac{F(x+h) - F(x) - hF_{(1)}(x)}{h^2} + \frac{1}{h} \int_0^h \frac{F(x+t) - F(x) - tF_{(1)}(x)}{t^2}\,dt.$$

Letting $h \to 0$, we have by the lemma

$$BD_1 f(x) = \lim_{h \to 0} \frac{1}{h} \int_0^h \frac{f(x+t) - BD_0 f(x)}{t}\,dt = \frac{1}{2}F_{(2)}(x) + \frac{1}{2}F_{(2)}(x) = F_{(2)}(x).$$

If $F_{(2)}(x) = \infty$, then for any $G > 0$ there is a $\delta > 0$ such that $\frac{2}{t^2}[F(x+t) - F(x) - tF_{(1)}(x)] > G$ for $|t| < \delta$ and, so if $|h| < \delta$, then $\frac{1}{h} \int_0^h \frac{F(x+t) - F(x) - tF_{(1)}(x)}{t^2/2}\,dt > G$, and so, letting $h \to 0$ in (2.7.8),

$$\liminf_{h \to 0} \frac{1}{h} \int_0^h \frac{f(x+t) - BD_0 f(x)}{t}\,dt \geq \infty + \frac{1}{2}G = \infty,$$

showing that $BD_1 f(x) = \infty$.

If $F_{(2)}(x) = -\infty$, the proof is similar.

To prove the inequality (2.7.6) in this case we consider only the last inequality and we may suppose that $\overline{F}_{(3)}(x) < \infty$, $\overline{F}_{(3)}(x) < M$, say. Since $F_{(2)}(x)$ is finite, we have by the above that $BD_1 f(x)$ is also finite. So, for $0 < \epsilon < h$, integrating by parts,

$$\int_\epsilon^h \frac{f(x+t) - BD_0 f(x) - tBD_1 f(x)}{t^2/2} \, dt$$

$$= \left. \frac{F(x+t) - F(x) - tBD_0 f(x) - (t^2/2)BD_1 f(x)}{t^2/2} \right|_\epsilon^h$$

$$+ \frac{4}{3!} \int_\epsilon^h \frac{F(x+t) - F(x) - tBD_0 f(x) - (t^2/2)BD_1 f(x)}{t^3/3!} \, dt. \quad (2.7.9)$$

Since $\overline{F}_{(3)}(x) < M$, there is a $\delta > 0$ such that the integrand in the last term on the right-hand side of (2.7.9) is less than M if $|t| < \delta$ and, so, if $0 < \epsilon < h < \delta$, the integral in the last term on the right-hand side of (2.7.9) does not exceed $M(h - \epsilon)$. Hence, from (2.7.9)

$$\frac{1}{h} \int_\epsilon^h \frac{f(x+t) - BD_0 f(x) - tBD_1 f(x)}{t^2/2} \, dt$$

$$\leq \left(\frac{1}{h}\right) \left. \frac{F(x+t) - F(x) - tBD_0 f(x) - (t^2/2)BD_1 f(x)}{t^2/2} \right|_\epsilon^h + \frac{2}{3} \frac{M(h - \epsilon)}{h}.$$

Letting $\epsilon \to 0$ and then $h \to 0+$ we have $\overline{BD}_2^+ f(x) \leq M$.

In a similar manner $\overline{BD}_2^- f(x) \leq M$ and so $\overline{BD}_2 f(x) \leq M$. Since M was arbitrary, this shows that $\overline{BD}_2 f(x) \leq \overline{F}_{(3)}(x)$, completing the proof of the theorem in the case $r = 1$.

Now assume the result to be known for $r = m$ and let $r = m + 1$.

Then, since $F_{(m+2)}(x)$ exists, $F_{(m+1)}(x)$ exists finitely and by the induction hypothesis, $BD_m f(x)$ exists and $BD_i f(x) = F_{(i+1)}(x)$, $0 \leq i \leq m$. Then writing $Q(t) = \sum_{i=0}^m \frac{t^i}{i!} BD_i f(x)$ as in (1.10.3) of Chapter I we have, integrating by parts,

$$\int_\epsilon^h \frac{f(x+t) - Q(t)}{t^{m+1}} \, dt = \left. \frac{F(x+t) - F(x) - \sum_{i=0}^m \frac{t^{i+1}}{(i+1)!} F_{(i+1)}(x)}{t^{m+1}} \right|_\epsilon^h$$

$$+ (m+1) \int_\epsilon^h \frac{F(x+t) - F(x) - \sum_{i=0}^m \frac{t^{i+1}}{(i+1)!} F_{(i+1)}(x)}{t^{m+2}} \, dt. \quad (2.7.10)$$

If $F_{(m+2)}(x)$ is finite then the right-hand side of (2.7.10) tends to a limit as $\epsilon \to 0$, and so, letting $\epsilon \to 0$ and then dividing by h, we get from (2.7.10)

$$\frac{1}{h}\int_0^h \frac{f(x+t) - Q(t)}{t^{m+1}}\,dt = \frac{F(x+h) - F(x) - \sum_{i=1}^{m+1}\frac{h^i}{i!}F_{(i)}(x)}{h^{m+2}}$$

$$+ \frac{m+1}{h}\int_0^h \frac{F(x+t) - F(x) - \sum_{i=1}^{m+1}\frac{t^i}{i!}F_{(i)}(x)}{t^{m+2}}\,dt, \qquad (2.7.11)$$

and so letting $h \to 0$, we have by of Lemma 2.7.3 that

$$BD_{m+1}f(x) = \frac{1}{m+2}F_{(m+2)}(x) + \frac{m+1}{m+2}F_{(m+2)}(x) = F_{(m+2)}(x).$$

If $F_{(m+2)}(x) = \infty$, then for any $G > 0$ there is a $\delta > 0$ such that

$$\frac{(m+2)!}{t^{m+2}}\left[F(x+t) - F(x) - \sum_{i=1}^{m+1}\frac{t^i}{i!}F_{(i)}(x)\right] > G \text{ for } |t| < \delta,$$

and so

$$\frac{1}{h}\int_0^h \frac{(m+2)!}{t^{m+2}}\left[F(x+t) - F(x) - \sum_{i=1}^{m+1}\frac{t^i}{i!}F_{(i)}(x)\right]\,dt > G \text{ for } |h| < \delta,$$

and hence, letting $h \to 0$ in (2.7.11),

$$\liminf_{h\to 0} \frac{1}{h}\int_0^h \frac{f(x+t) - Q(t)}{t^{m+1}}\,dt \geq \infty + \frac{m+1}{(m+2)!}G = \infty.$$

This shows that $BD_{m+1}f(x) = \infty$.

If $F_{(m+2)}(x) = -\infty$, the proof is similar.

This completes the proof of the first part of the theorem. For the second part, we can suppose that $\overline{F}_{(m+3)}(x) < M < \infty$. Then $F_{(m+2)}(x)$ exists finitely and so by the first part $BD_{m+1}f(x)$ exists and $BD_if(x) = F_{(i+1)}(x)$, $0 \leq i \leq m+1$. So, for $0 < \epsilon < h$, integrating by parts,

$$\int_\epsilon^h \frac{f(x+t) - \sum_{i=0}^{m+1}\frac{t^i}{i!}BD_if(x)}{t^{m+2}}\,dt = \qquad (2.7.12)$$

$$\left.\frac{F(x+t) - F(x) - \sum_{i=0}^{m+1}\frac{t^{i+1}}{(i+1)!}F_{(i+1)}(x)}{t^{m+2}}\right|_\epsilon^h$$

$$+ (m+2)\int_\epsilon^h \frac{F(x+t) - F(x) - \sum_{i=0}^{m+1}\frac{t^{i+1}}{(i+1)!}F_{(i+1)}(x)}{t^{m+3}}\,dt.$$

Since $\overline{F}_{(m+3)}(x) < M$, there is a $\delta > 0$ such that

$$\frac{F(x+t) - F(x) - \sum_{i=1}^{m+2}\frac{t^i}{i!}F_{(i)}(x)}{t^{m+3}/(m+3)!} < M \text{ for } |t| < \delta.$$

So, if $0 < \epsilon < h < \delta$, then

$$\int_\epsilon^h \frac{F(x+t) - F(x) - \sum_{i=1}^{m+2} \frac{t^i}{i!} F_{(i)}(x)}{t^{m+3}/(m+3)!} \, dt < M(h - \epsilon)$$

and, hence, from (2.7.12)

$$\frac{1}{h} \int_\epsilon^h \frac{f(x+t) - \sum_{i=0}^{m+1} \frac{t^i}{i!} BD_i f(x)}{t^{m+2}} \, dt$$

$$\leq \frac{1}{h} \left[\frac{F(x+t) - F(x) - \sum_{i=0}^{m+1} \frac{t^{i+1}}{(i+1)!} F_{(i+1)}(x)}{t^{m+2}} \Big|_\epsilon^h \right] + \frac{m+2}{(m+3)!} \frac{M(h-\epsilon)}{h}$$

$$< \frac{M}{(m+3)!} - \frac{F(x+\epsilon) - F(x) - \sum_{i=1}^{m+2} \frac{\epsilon^i}{i!} F_{(i)}(x)}{h \epsilon^{m+2}} + \frac{m+2}{(m+3)!} \frac{M(h-\epsilon)}{h}.$$

Letting $\epsilon \to 0+$ first and then $h \to 0+$, we have

$$\overline{BD}^+_{m+2} f(x) \leq \frac{M}{m+3} + \frac{(m+2)M}{m+3} = M.$$

Similarly, $\overline{BD}^-_{m+2} f(x) \leq M$ and so $\overline{BD}_{m+2} f(x) \leq M$. Since M is arbitrary, this proves that $\overline{BD}_{m+2} f(x) \leq \overline{F}_{(m+3)}(x)$. □

Theorem 2.7.4 *Let f be special Denjoy integrable in some neighbourhood of x and let F be its indefinite integral in that neighbourhood. If $BD_r f(x)$ exists finitely, then $F_{(r+1)}(x)$ exists and equals $BD_r f(x)$. If, moreover,*

$$\int_0^h \frac{f(x+t) - \sum_{i=0}^r \frac{t^i}{i!} BD_i f(x)}{t^{r+1}} \, dt = \lim_{\epsilon \to 0} \int_\epsilon^h \frac{f(x+t) - \sum_{i=0}^r \frac{t^i}{i!} BD_i f(x)}{t^{r+1}} \, dt$$

exists, then

$$(r+2)\underline{BD}_{r+1} f(x) - (r+1)\overline{BD}_{r+1} f(x) \leq \underline{F}_{(r+2)}(x)$$
$$\leq \overline{F}_{(r+2)}(x) \leq (r+2)\overline{BD}_{r+1} f(x) - (r+1)\underline{BD}_{r+1} f(x), \quad (2.7.13)$$

provided the two extreme members on the left and right of (2.7.13) are well defined.

□ The last condition excludes the possibility that $BD_{r+1} f(x)$ exists and is infinite. Further note that (2.7.13) is obvious if either $\overline{BD}_{r+1} f(x) = \infty$ and $\underline{BD}_{r+1} f(x) < \infty$ or if $\underline{BD}_{r+1} f(x) = -\infty$ and $\overline{BD}_{r+1} f(x) > -\infty$.

First, we prove the result when $r = 1$, that is, if $BD_1 f(x)$ exists finitely. Let

$$\phi(h) = \int_0^h \frac{f(x+t) - BD_0 f(x)}{t} \, dt,$$

the integral existing from the definition of $BD_1 f(x)$.

Since ϕ is a special Denjoy integral, $\phi'(t) = \frac{1}{t}\left[f(x+t) - BD_0 f(x)\right]$ almost everywhere, and so $t\phi'(t) = \left[f(x+t) - BD_0 f(x)\right]$ almost everywhere. Hence,

$$\int_0^h t\phi'(t)\, dt = \int_0^h \left[f(x+t) - BD_0 f(x)\right] dt = F(x+h) - F(x) - hBD_0 f(x).$$
(2.7.14)

Also, since ϕ is a special Denjoy integral, $\int_0^h \phi' = \phi(h) - \phi(0) = \phi(h)$, and so from (2.7.14), on integrating by parts,

$$F(x+h) - F(x) - hBD_0 f(x) = h\phi(h) - \int_0^h \phi.$$

Hence,

$$\frac{F(x+h) - F(x) - hBD_0 f(x)}{h^2/2} = 2\frac{\phi(h)}{h} - \frac{2}{h^2}\int_0^h \phi.$$
(2.7.15)

Since $BD_1 f(x)$ exists finitely, $\frac{\phi(h)}{h} \to BD_1 f(x)$ and $\frac{2}{h^2}\int_0^h \phi \to BD_1 f(x)$ as $h \to 0$, so from (2.7.15), letting $h \to 0$, $F_{(2)}(x)$ exists with $F_{(1)}(x) = BD_0 f(x)$ and $F_{(2)}(x) = 2BD_1 f(x) - BD_1 f(x) = BD_1 f(x)$. This gives the first part in this case.

For the second part, let

$$\psi(h) = \int_0^h \frac{f(x+t) - BD_0 f(x) - tBD_1 f(x)}{t^2}\, dt$$

exist. Then, as above, $t^2\psi'(t) = f(x+t) - BD_0 f(x) - tBD_1 f(x)$ almost everywhere and, so, on integrating,

$$\int_0^h t^2\psi'(t)\, dt = F(x+h) - F(x) - hBD_0 f(x) - \frac{h^2}{2}BD_1 f(x).$$

Hence, integrating by parts and noting that $F_{(1)}(x) = BD_0 f(x)$ and $F_{(2)}(x) = BD_1 f(x)$,

$$F(x+h) - F(x) - hF_{(1)}(x) - \frac{h^2}{2}F_{(2)}(x) = h^2\psi(h) - 2\int_0^h t\psi(t)\, dt.$$

Applying the mean value theorem,

$$\frac{F(x+h) - F(x) - hF_{(1)}(x) - \frac{h^2}{2}F_{(2)}(x)}{h^3/3!} = 3!\frac{\psi(h)}{h} - \frac{3!2}{h^3}\int_0^h t\psi(t)\, dt$$

$$= 3!\frac{\psi(h)}{h} - \frac{3!2}{3}\frac{\xi\psi(\xi)}{\xi^2}$$

for some ξ, $0 < \xi < h$. Letting $h \to 0$,

$$\overline{F}_{(3)}(x) \leq 3! \limsup_{h \to 0} \frac{\psi(h)}{h} - 2!2 \liminf_{h \to 0} \frac{\psi(\xi)}{\xi}$$

$$\leq 3\overline{BD}_2 f(x) - 2!2 \liminf_{h \to 0} \frac{\psi(h)}{h}$$

$$= 3\overline{BD}_2 f(x) - 2\underline{BD}_2 f(x).$$

In a similar manner, we can obtain $\underline{F}_{(3)}(x) \geq 3\underline{BD}_2 f(x) - 2\overline{BD}_2 f(x)$, completing the proof of the theorem for this case, $r = 1$.

Now, assume the result to be known for $r = m$ and let $r = m + 1$. Let $BD_{m+1}f(x)$ exist finitely. Then $BD_i f(x)$ exists, $0 \leq i \leq m$. Let

$$\phi(h) = \int_0^h \frac{f(x+t) - \sum_{i=0}^m \frac{t^i}{i!} BD_i f(x)}{t^{m+1}} \, dt,$$

the integral existing by the definition of $BD_{m+1}f(x)$. Since ϕ is a Denjoy integral, $\phi'(t) = \dfrac{f(x+t) - \sum_{i=0}^m \frac{t^i}{i!} BD_i f(x)}{t^{m+1}}$ almost everywhere, which gives

$$\int_0^h t^{m+1} \phi'(t) \, dt = F(x+h) - F(x) - \sum_{i=0}^m \frac{h^{i+1}}{(i+1)!} BD_i f(x). \qquad (2.7.16)$$

By the induction hypothesis, $F_{(i+1)}(x)$ exists and is equal to $BD_i f(x), 0 \leq i \leq m$, and, so, from (2.7.16) on integrating by parts,

$$F(x+h) - F(x) - \sum_{i=1}^{m+1} \frac{h^i}{i!} F_{(i)}(x) = h^{m+1}\phi(h) - (m+1)\int_0^h t^m \phi(t) \, dt.$$

Hence,

$$\frac{F(x+h) - F(x) - \sum_{i=1}^{m+1} \frac{h^i}{i!} F_{(i)}(x)}{h^{m+2}/(m+2)!} = (m+2)! \left[\frac{\phi(h)}{h} - \frac{m+1}{h^{m+2}} \int_0^h t^m \phi(t) \, dt \right].$$

$$(2.7.17)$$

Since $(m+1)!(\phi(h)/h) \to BD_{m+1}f(x)$ as $h \to 0$, we have from (2.7.17), letting $h \to 0$,

$$F_{(m+2)}(x) = (m+2)BD_{m+1}f(x) - (m+1)BD_{m+1}f(x) = BD_{m+1}f(x).$$

$$(2.7.18)$$

For the second part, let

$$\psi(h) = \int_0^h \frac{f(x+t) - \sum_{i=0}^{m+1} \frac{t^i}{i!} BD_i f(x)}{t^{m+2}} \, dt$$

exist. Then, as above, $t^{m+2}\psi'(t) = f(x+t) - \sum_{i=0}^{m+1} \frac{t^i}{i!} BD_i f(x)$ almost everywhere and so

$$\int_0^h t^{m+2} \psi'(t) \, dt = F(x+h) - F(x) - \sum_{i=0}^{m+1} \frac{h^{i+1}}{(i+1)!} BD_i f(x). \qquad (2.7.19)$$

As $BD_{m+1}f(x)$ exists finitely, by the induction hypothesis, $BD_if(x) = F_{(i+1)}(x)$, $0 \le i \le m$. Also, from (2.7.18), $F_{(m+2)}(x) = BD_{m+1}f(x)$ and so, from (2.7.19) on integrating by parts,

$$F(x+h) - F(x) - \sum_{i=1}^{m+2} \frac{h^i}{i!} F_{(i)}(x) = h^{m+2}\psi(h) - (m+2) \int_0^h t^{m+1}\psi(t)\,dt.$$

Hence, applying the mean value theorem,

$$\frac{F(x+h) - F(x) - \sum_{i=1}^{m+2} \frac{h^i}{i!} F_{(i)}(x)}{h^{m+3}/(m+3)!}$$

$$= (m+3)! \Big[\frac{\psi(h)}{h} - \frac{(m+2)}{h^{m+3}} \int_0^h t^{m+1}\psi(t)\,dt \Big]$$

$$= (m+3)! \frac{\psi(h)}{h} - (m+2)!(m+2) \frac{\xi^{m+1}\psi(\xi)}{\xi^{m+2}},$$

for some ξ, $0 < \xi < h$. Letting $h \to 0$,

$$\overline{F}_{(m+3)}(x) \le (m+3)! \limsup_{h\to 0} \frac{\psi(h)}{h} - (m+2)!(m+2) \liminf_{h\to 0} \frac{\psi(\xi)}{\xi}$$

$$\le (m+3)\overline{BD}_{m+2}f(x) - (m+2)\underline{BD}_{m+2}f(x).$$

In a similar manner $\underline{F}_{(m+3)}(x) \ge (m+3)\underline{BD}_{m+2}f(x) - (m+2)\overline{BD}_{m+2}f(x)$, which completes the proof. ☐

Remark. All the results concerning upper and lower derivates proved above are also true for the corresponding unilateral upper and lower derivates.

2.8 Symmetric Borel and Symmetric de la Vallée Poussin Derivatives, SBD_kf and $f^{(s)}_{(k)}$

Theorem 2.8.1 *Let f be special Denjoy integrable in some neighbourhood of x. If $f^{(s)}_{(r)}(x)$ exists, then $SBD_rf(x)$ exists and the two derivatives are equal; the converse is not true. More generally, if $f^{(s)}_{(r-2)}(x)$ exists finitely, then*

$$\underline{f}^{(s)}_{(r)}(x) \le \underline{SBD}_rf(x) \le \overline{SBD}_rf(x) \le \overline{f}^{(s)}_{(r)}(x).$$

☐   *If $f^{(s)}_{(r)}(x)$ exists finitely, then* $\dfrac{f(x+t) + (-1)^r f(x-t)}{2} = P(t) + o(t^r)$

where

$$P(t) = \begin{cases} \sum_{i=0}^{r/2} \dfrac{t^{2i}}{(2i)!} f^{(s)}_{(2i)}(x), & \text{if } r \text{ is even,} \\[2ex] \sum_{i=0}^{(r-1)/2} \dfrac{t^{2i+1}}{(2i+1)!} f^{(s)}_{(2i+1)}(x) & \text{if } r \text{ is odd.} \end{cases}$$

Hence,

$$\frac{1}{h} \int_0^h \frac{1}{t^r} \left(\frac{f(x+t) + (-1)^r f(x-t)}{2} - P(t) \right) dt = o(1) \text{ as } h \to 0.$$

So, $SBD_r f(x)$ exists and $SBD_r f(x) = f_{(r)}^{(s)}(x)$.

If $f_{(r)}^{(s)}(x)$ is infinite, then by definition $f_{(i)}^{(s)}(x)$ exists finitely, $i = r - 2$, $r - 4, \cdots$, and so, by the above argument, $SBD_i f(x)$ exists and equals $f_{(i)}^{(s)}(x)$, $i = r - 2, r - 4, \cdots$.

Let $f_{(r)}^{(s)}(x) = \infty$. Then for any $M > 0$ there is a $\delta > 0$ such that $\varpi_r(f; x, t) > M$ for $0 < |t| < \delta$, where $\varpi_r(f; x, t)$ is defined in (2.7.5) of Chapter I. So,

$$\frac{1}{h} \int_\epsilon^h \varpi_r(f; x, t) \, dt > M \frac{h - \epsilon}{h}, \text{ for } 0 < \epsilon < h < \delta. \tag{2.8.1}$$

Since by the above argument, $SBD_i f(x)$ exists and equals $f_{(i)}^{(s)}(x)$, $i = r - 2$, $r - 4, \cdots$, $\varpi_r(f; x, t) = \overline{\varpi}_r(f; x, t)$, where $\overline{\varpi}_r(f; x, t)$ is defined as in (1.11.6) of Chapter I, and, hence, from (2.8.1) we have

$$\frac{1}{h} \int_\epsilon^h \overline{\varpi}_r(f; x, t) \, dt > M \frac{h - \epsilon}{h}, \text{ for } 0 < \epsilon < h < \delta.$$

Letting $\epsilon \to 0+$ first and then $h \to 0+$, we get that $\underline{SBD_r} f(x) \geq M$ and, so, since M is arbitrary, $\underline{SBD_r} f(x) = \infty$; this proves that $SBD_r f(x) = f_{(r)}^{(s)}(x)$.

If $f_{(r)}^{(s)}(x) = -\infty$, the argument is similar.

Regarding the converse, let r be a fixed positive integer and consider the functions:

$$f(x) = \begin{cases} \cos x, \\ x^r \sin(1/x^{r-1}); \end{cases} \qquad g(x) = \begin{cases} \sin x, & \text{if } x \text{ is rational,} \\ x^r \sin(1/x^{r-1}), & \text{if } x \text{ is irrational.} \end{cases}$$

Since $f_{(2)}^{(s)}(0)$ does not exist, then $f_{(r)}^{(s)}(0)$ does not exist if r is even, and since $g_{(1)}^{(s)}(0)$ does not exist, then $g_{(r)}^{(s)}(0)$ does not exist if r is odd. However, for any r,

$$\frac{1}{h} \int_0^h \frac{f(t) + (-1)^r f(-t)}{2t^r} \, dt = \frac{1}{h} \int_0^h \frac{g(t) + (-1)^r g(-t)}{2t^r} \, dt = I, \text{ say.}$$

Further,

$$I = \begin{cases} 0, & \text{if } r \text{ is even,} \\ \frac{1}{h} \int_0^h \sin(1/t^{r-1}) \, dt, & \text{if } r \text{ is odd.} \end{cases}$$

Hence, proceeding as in Theorem 2.7.1, $I = o(1)$ as $h \to 0$, $\big($see (2.7.4)$\big)$, and so whether r is even or odd both $SBD_r f(0)$ and $SBD_r g(0)$ exist.

To prove the converse, take f or g according as r is even or odd.

We now consider the last part of the theorem and consider the last inequality; the first can be treated in a similar manner. We may suppose that $\overline{f}^{(s)}_{(r)}(x) < \infty$, and let $\overline{f}^{(s)}_{(r)}(x) < M < \infty$. Then there is a $\delta > 0$ such that $\varpi_r(f; x, t) < M$ for $0 < |t| < \delta$. Hence,

$$\frac{1}{h}\int_\epsilon^h \varpi_r(f; x, t)\, dt \le M\frac{h - \epsilon}{h}, \quad \text{for } 0 < \epsilon < h < \delta. \tag{2.8.2}$$

Since $f^{(s)}_{(r-2)}(x)$ exists finitely, then by the first part $SBD_{r-2}f(x)$ exists finitely and is equal to $f^{(s)}_{(r-2)}(x)$ and thus $f^{(s)}_{(i)}(x) = SBD_i f(x), i = r - 2, r - 4, \cdots$, from which if follows that $\varpi_r(f; x.t) = \overline{\varpi}_r(f; x.t)$ and, so, letting $\epsilon \to 0$ and then $h \to 0$ in (2.8.2), we have $\overline{SBD}_r f(x) \le M$. Since M was arbitrary, this completes the proof that $\overline{SBD}_r f(x) \le \overline{f}^{(s)}_{(r)}(x)$. $\qquad\square$

Theorem 2.8.2 *Let f be special Denjoy integrable in some neighbourhood of x and let F be its indefinite integral in that neighbourhood. If $SBD_r f(x)$ exists finitely, then $F^{(s)}_{(r+1)}(x)$ exists with value $SBD_r f(x)$. Moreover, if $\int_0^h \overline{\varpi}_{r+2}(f; x.t)\, dt$ exists, then*

$$(r + 3)\underline{SBD}_{r+2}f(x) - (r + 2)\overline{SBD}_{r+2}f(x) \le \underline{F}^{(s)}_{(r+3)}(x) \le \overline{F}^{(s)}_{(r+3)}(x)$$
$$\le (r + 3)\overline{SBD}_{r+2}f(x) - (r + 2)\underline{SBD}_{r+2}f(x) \tag{2.8.3}$$

provided the two extreme members in the left and right of the inequality are properly defined.

$\square$ If either $\overline{SBD}_{r+2}f(x) = \infty$ and $\underline{SBD}_{r+2}f(x) < \infty$ or if $\underline{SBD}_{r+2}f(x) = -\infty$ and $\overline{SBD}_{r+2}f(x) > -\infty$, then (2.8.3) is obvious. Also, if $SBD_{r+2}f(x)$ exists, it must be finite; for otherwise the extreme members become undefined.

We first prove the case $r = 1$.

Let $SBD_1 f(x)$ exist finitely and write

$$\phi(h) = \int_0^h \frac{f(x + t) - f(x - t)}{2t}\, dt;$$

this integral exists by the definition of $SBD_1 f(x)$. Since ϕ is a special Denjoy integral, $\phi'(t) = \frac{f(x + t) - f(x - t)}{2t}$ almost everywhere and $\int_0^h \phi' = \phi(h) - \phi(0) = \phi(h)$. Since $2t\phi'(t) = f(x + t) - f(x - t)$ almost everywhere,

$$2\int_0^h t\phi'(t)\, dt = \int_0^h \left[f(x + t) - f(x - t)\right] dt = F(x + h) - 2F(x) + F(x + h).$$

On integrating by parts and dividing by h^2, this gives

$$\frac{F(x + h) - 2F(x) + F(x + h)}{h^2} = \frac{2}{h^2}\left[h\phi(h) - \int_0^h \phi\right] = 2\frac{\phi(h)}{h} - \frac{2}{h^2}\int_0^h \phi. \tag{2.8.4}$$

Since $SBD_1f(x)$ exists $\dfrac{\phi(h)}{h}$ and $\dfrac{2}{h^2}\displaystyle\int_0^h \phi \to SBD_1f(x)$ as $h \to 0$. Hence, from (2.8.4) on letting $h \to 0$, we obtain $F_{(2)}^{(s)}(x) = SBD_1f(x)$, giving the first part of the theorem when $r = 1$. For the second part of the theorem in this case, let

$$\psi(h) = \int_0^h \overline{\varpi}_3(f; x, t)\, \mathrm{d}t$$

exist. Then, as above, $\psi'(t) = \overline{\varpi}_3(f; x, t)$ almost everywhere and so

$$\frac{t^3}{3!}\psi'(t) = \frac{f(x+t) - f(x-t)}{2} - t SBD_1 f(x)$$

almost everywhere and, hence,

$$
\begin{aligned}
\frac{1}{3!}\int_0^h t^3\psi'(t)\, \mathrm{d}t &= \int_0^h \left[\frac{f(x+t) - f(x-t)}{2} - t SBD_1 f(x)\right] \mathrm{d}t \\
&= \frac{F(x+h) + F(x-h)}{2} - F(x) - \frac{h^2}{2} SBD_1 f(x) \\
&= \frac{F(x+h) + F(x-h)}{2} - F(x) - \frac{h^2}{2} F_{(2)}^{(s)}(x).
\end{aligned}
$$

Integrating by parts and dividing by $h^4/4!$ this becomes

$$
\begin{aligned}
\frac{4!}{h^4}\Bigg[\frac{F(x+h) + F(x-h)}{2} - F(x) \quad &- \frac{h^2}{2} F_{(2)}^{(s)}(x)\Bigg] \\
= \tfrac{4}{h^4}\left[h^3\psi(h) - 3\int_0^h t^2\psi(t)\, \mathrm{d}t\right] &= 4\left[\tfrac{\psi(h)}{h} - \tfrac{3}{h^4}\int_0^h t^2\psi(t)\, \mathrm{d}t\right]. \quad (2.8.5)
\end{aligned}
$$

By the mean value theorem there is a ξ, $0 < \xi < h$, such that

$$\frac{1}{h^4}\int_0^h t^2\psi(t)\, \mathrm{d}t = \frac{1}{4\xi^3}\xi^2\psi(\xi) = \frac{1}{4}\frac{\psi(\xi)}{\xi}, \qquad (2.8.6)$$

and so letting $h \to 0$, we have from (2.8.5) and (2.8.6)

$$\overline{F}_{(4)}^{(s)}(x) \le 4\limsup_{h \to 0} \frac{\psi(h)}{h} - 3\liminf_{h \to 0}\frac{\psi(h)}{h} \le 4\overline{SBD}_3 f(x) - 3\underline{SBD}_3 f(x),$$

and

$$\underline{F}_{(4)}^{(s)}(x) \ge 4\liminf_{h \to 0}\frac{\psi(h)}{h} - 3\limsup_{h \to 0}\frac{\psi(h)}{h} \ge 4\underline{SBD}_3 f(x) - 3\overline{SBD}_3 f(x).$$

This completes the discussion of the case $r = 1$.

Let us now suppose that the result is true for $r = 2m - 1$ and prove it for $r = 2m + 1$.

Assume that $SBD_{2m+1}f(x)$ exists. Then $SBD_{2i+1}f(x)$, $1 = 0, 1, \cdots, m-1$ exists. Let

$$\phi(h) = \int_0^h \overline{\varpi}_{2m+1}(f; x, t)\, dt,$$

the integral existing by the definition of $SBD_{2m+1}f(x)$. Then, as above,

$$\frac{1}{(2m+1)!} \int_0^h t^{2m+1} \phi'(t)\, dt$$

$$= \int_0^h \left[\frac{f(x+t) - f(x-t)}{2} - \sum_{i=0}^{m-1} \frac{t^{2i+1}}{(2i+1)!} SBD_{2i+1}f(x) \right] dt \quad (2.8.7)$$

$$= \frac{F(x+h) + F(x-h)}{2} - F(x) - \sum_{i=0}^{m-1} \frac{h^{2i+2}}{(2i+2)!} SBD_{2i+1}f(x).$$

Since $SBD_{2i+1}f(x)$ exists finitely, $i = 0, 1, \cdots, m-1$, by the induction hypothesis, $F_{(2i+2)}^{(s)}(x)$ exists and is equal to $SBD_{2i+1}f(x)$, $i = 0, 1, \cdots, m-1$ and so (2.8.7) becomes

$$\frac{F(x+h) + F(x-h)}{2} - F(x) - \sum_{i=1}^{m} \frac{h^{2i}}{(2i)!} F_{(2i)}^{(s)}(x)$$

$$= \frac{1}{(2m+1)!} \int_0^h t^{2m+1} \phi'(t)\, dt$$

$$= \frac{1}{(2m+1)!} \left[h^{2m+1}\phi(h) - (2m+1) \int_0^h t^{2m}\phi(t)\, dt \right].$$

So,

$$\frac{(2m+2)!}{h^{2m+2}} \left[\frac{F(x+h) + F(x-h)}{2} - F(x) - \sum_{i=1}^{m} \frac{h^{2i}}{(2i)!} F_{(2i)}^{(s)}(x) \right]$$

$$= (2m+2)\frac{\phi(h)}{h} - \frac{(2m+1)(2m+2)}{h^{2m+2}} \int_0^h t^{2m}\phi(t)\, dt. \quad (2.8.8)$$

Since $\dfrac{\phi(h)}{h} \to SBD_{2m+1}f(x)$ and $\dfrac{2m+2}{h^{2m+2}} \displaystyle\int_0^h t^{2m}\phi(t)\, dt \to SBD_{2m+1}f(x)$ as $h \to 0$, letting $h \to 0$ in (2.8.8), we obtain

$$F_{(2m+2)}^{(s)}(x) = (2m+2)SBD_{2m+1}f(x) - (2m+1)SBD_{2m+1}f(x) = SBD_{2m+1}f(x),$$

proving the first part of of the theorem for $r = 2m + 1$.
    For the second part, let

$$\psi(h) = \int_0^h \overline{\varpi}_{2m+3}(f; x, t)\, d$$

exist. Then as above,

$$\frac{1}{(2m+3)!} \int_0^h t^{2m+3} \psi'(t)\,dt$$

$$= \int_0^h \left[\frac{f(x+t)-f(x-t)}{2} - \sum_{i=0}^{m} \frac{t^{2i+1}}{(2i+1)!} SBD_{2i+1}f(x) \right] dt \quad (2.8.9)$$

$$= \frac{F(x+h)+F(x-h)}{2} - F(x) - \sum_{i=0}^{m} \frac{h^{2i+2}}{(2i+2)!} SBD_{2i+1}f(x).$$

By the induction hypothesis $SBD_{2i+1}f(x) = F_{(2i+2)}^{(s)}(x)$, $i = 0, 1, \cdots, m-1$, and, so, from (2.8.9)

$$\frac{1}{(2m+3)!} \int_0^h t^{2m+3} \psi'(t)\,dt$$

$$= \frac{F(x+h)+F(x-h)}{2} - F(x) - \sum_{i=1}^{m+1} \frac{h^{2i}}{(2i)!} F_{(2i)}^{(s)}(x). \quad (2.8.10)$$

Also, integrating by parts and applying the mean value theorem,

$$\frac{1}{h^{2m+4}} \int_0^h t^{2m+3} \psi'(t)\,dt = \frac{\psi(h)}{h} - \frac{2m+3}{h^{2m+4}} \int_0^h t^{2m+2} \psi(t)\,dt$$

$$= \frac{\psi(h)}{h} - \frac{2m+3}{2m+4} \frac{\psi(\xi)}{\xi}, \quad (2.8.11)$$

where $0 < \xi < h$. So, multiplying both sides of (2.8.10) by $(2m+4)!/h^{2m+4}$ and using (2.8.11), we have

$$\varpi_{2m+4}(F;x,t) = (2m+4)\frac{\psi(h)}{h} - (2m+3)\frac{\psi(\xi)}{\xi}.$$

Letting $h \to 0$, we then have

$$F_{(2m+4)}^{(s)}(x) \leq (2m+4)\limsup_{h\to 0}\frac{\psi(h)}{h} - (2m+3)\liminf_{h\to 0}\frac{\psi(\xi)}{\xi}$$

$$\leq (2m+4)\overline{SBD}_{2m+3}f(x) - (2m+3)\underline{SBD}_{2m+3}f(x),$$

and similarly

$$\underline{F}_{(2m+4)}^{(s)}(x) \geq (2m+4)\underline{SBD}_{2m+3}f(x) - (2m+3)\overline{SBD}_{2m+3}f(x),$$

completing the proof for $r = 2m+1$ and so the result is proved by induction for all odd values of r.

The proof for even values of r is similar. $\qquad\square$

Theorem 2.8.3 *Let f be special Denjoy integrable in some neighbourhood of x. If f is smooth of order r at x, then it is Borel smooth of order r at x. More generally, if $f^{(s)}_{(r-2)}(x)$ exists finitely, then*

$$\underline{S}_r f(x) \leq \underline{BS}_r f(x) \leq \overline{BS}_r f(x) \leq \overline{S}_r f(x), \qquad (2.8.12)$$

where $\underline{S}_r f(x), [\overline{S}_r f(x)]$, and $\underline{BS}_r f(x), [\overline{BS}_r f(x)]$, are, respectively, the index of lower, [upper], smoothness and Borel smoothness of f of order r at x.

[For the various definitions, see §6.2 and §11.2 of Chapter I.]

□   We prove (2.8.12), which implies the first part. Since $f^{(s)}_{(r-2)}(x)$ exists finitely, then by the first part of Theorem 2.8.1 $SBD_i f(x)$ exists and is equal to $f^{(s)}_{(i)}(x)$, $i = r-2, r-4, \cdots$. Hence, $\varpi_r(f; x, t) = \overline{\varpi}_r(f; x, t)$.

To prove the last inequality of (2.8.12), we may suppose that $\overline{S}_r f(x) < \infty$ and then choose an M such that $\overline{S}_r f(x) < M$. Since $\overline{S}_r f(x) = \limsup_{t \to 0} t \varpi_r(f; x, t)$, there is a $\delta > 0$ such that $t \varpi_r(f; x, t) < M$ for $0 < t < \delta$ and, so, since $\varpi_r(f; x, t) = \overline{\varpi}_r(f; x, t)$, we have

$$\frac{1}{h} \int_\epsilon^h t \overline{\varpi}_r(f; x, t) \, dt \leq M \frac{h - \epsilon}{h}, \text{ for } 0 < \epsilon < h < \delta.$$

Letting $\epsilon \to 0$ and then $h \to 0$, we get that $\limsup_{h \to 0} \dfrac{1}{h} \int_\epsilon^h t \overline{\varpi}_r(f; x, t) \, dt \leq M$. Since M was arbitrary, the last inequality of (2.8.12) is proved.

The proof of the first inequality of (2.8.12) is similar.   □

The converse of the first part of Theorem 2.8.3 is not true and to see this we first give the following lemma that will be of use in the sequel.

Lemma 2.8.4 *If ϕ is special Denjoy integrable in some neighbourhood of 0 and if $\int_0^h \phi = o(h^\alpha)$ as $h \to 0$, then $\int_0^h t\phi(t) \, dt = o(h^{\alpha+1})$ as $h \to 0$.*

□   Integrating by parts,

$$\int_0^h t\phi(t) \, dt = h \int_0^h \phi - \int_0^h \int_0^t \phi(\xi) \, d\xi \, dt = o(h^{\alpha+1}) \text{ as } h \to 0.$$

□Now consider the functions f and g in Theorem 2.8.1 according as r is even or odd and apply Lemma 2.8.4 above to show that as $h \to 0$,

$$\frac{1}{h} \int_0^h \frac{1}{t^{r-1}} \frac{f(t) + (-1)^r f(-t)}{2} \, dt = \frac{1}{h} \int_0^h \frac{1}{t^{r-1}} \frac{g(t) + (-1)^r g(-t)}{2} \, dt = o(1).$$

This proves that f or g is Borel smooth at 0 of order r. However, these functions are not smooth of order r at 0 since $f^{(s)}_{(2)}(0)$, $g^{(s)}_{(2)}(0)$ do not exist.

Theorem 2.8.5 *Let f be special Denjoy integrable in some neighbourhood of x and let F be its indefinite integral in that neighbourhood. If f is Borel smooth at x of order r, $r \geq 2$, then F is smooth of order $r + 1$ at x. More generally, if $SBD_{r-2}f(x)$ exists finitely and if $\int_0^h t\overline{\omega}_r(f; x, t)\, dt$ exists, then*

$$(r+1)\underline{BS}_r f(x) - \frac{r^2 - 1}{r}\overline{BS}_r f(x) \leq \underline{S}_{r+1}F(x)$$

$$\leq \overline{S}_{r+1}F(x) \leq (r+1)\overline{BS}_r f(x) - \frac{r^2 - 1}{r}\underline{BS}_r f(x). \qquad (2.8.13)$$

☐   We prove (2.8.13) from which the first part of the theorem follows. We first consider the case $r = 2$.
Let

$$\psi(h) = \int_0^h t\overline{\omega}_2(f; x, t)\, dt \qquad (2.8.14)$$

exist. Then, as in Theorem 2.8.2, $\psi'(t) = t\overline{\omega}_2(f; x, t)$ almost everywhere and so

$$\int_0^h t\psi'(t)\, dt = 2\int_0^h \left[\frac{f(x+t)+f(x-t)}{2} - SBD_0 f(x)\right] dt$$

$$= F(x+h) - F(x-h) - 2hSBD_0 f(x).$$

On integrating by parts, this gives

$$F(x+h) - F(x-h) - 2hSBD_0 f(x) = h\psi(h) - \int_0^h \psi. \qquad (2.8.15)$$

By hypothesis $\psi(h) \to 0$ as $h \to 0$ and so $\frac{1}{h}\int_0^h \psi \to 0$ as $h \to 0$. Hence, dividing both sides of (2.8.15) by $2h$ and then letting $h \to 0$, we conclude that $F_{(1)}^{(s)}(x)$ exists with value $SBD_0 f(x)$. So, from (2.8.15), we have by the mean value theorem that

$$\frac{3!}{h^2}\left[\frac{F(x+h) - F(x-h)}{2} - hF_{(1)}^{(s)}(x)\right] = \frac{3\psi(h)}{h} - \frac{3}{h^2}\int_0^h \psi$$

$$= \frac{3\psi(h)}{h} - \frac{3\psi(\xi)}{2\xi}, \qquad (2.8.16)$$

where $0 < \xi < h$.
Since $\frac{\psi(h)}{h} = \frac{1}{h}\int_0^h t\overline{\omega}_2(f; x, t)\, dt$, we have from (2.8.16), on letting $h \to 0$, that

$$\overline{S}_3 F(x) \leq 3\overline{BS}_2 f(x) - \frac{3}{2}\underline{BS}_2 f(x), \text{ and } \underline{S}_3 F(x) \geq 3\underline{BS}_2 f(x) - \frac{3}{2}\overline{BS}_2 f(x),$$

proving the result in the case $r = 2$.

Suppose now that $r > 2$. We prove the case of even r; the case of odd r can be handled in a similar manner.

Let $r = 2m$ and suppose $SBD_{2m-2}f(x)$ exists finitely and that

$$\psi(h) = \int_0^h t\overline{\varpi}_{2m}(f; x, t)\, dt \qquad (2.8.17)$$

exists. Then, as above, (2.8.17) gives

$$\int_0^h t^{2m-1}\psi'(t)\, dt = (2m)! \int_0^h \left[\frac{f(x+t) + f(x-t)}{2} - \sum_{i=0}^{m-1} \frac{t^{2i}}{(2i)!} SBD_{2i}f(x) \right] dt$$

$$= (2m)! \left[\frac{F(x+h) - F(x-h)}{2} - \sum_{i=0}^{m-1} \frac{h^{2i+1}}{(2i+1)!} SBD_{2i}f(x) \right].$$

$$(2.8.18)$$

By Theorem 2.8.2, $F_{(2i-1)}^{(s)}(x)$ exists and equals $SBD_{2i-2}f(x)$, $i = 1, 2, \cdots, m$, and so from (2.8.18) on integrating by parts

$$(2m)! \left[\frac{F(x+h) - F(x-h)}{2} - \sum_{i=0}^{m-1} \frac{h^{2i+1}}{(2i+1)!} F_{(2i+1)}^{(s)}(x) \right]$$

$$= h^{2m-1}\psi(h) - (2m-1) \int_0^h t^{2m-2}\psi(t)\, dt,$$

and, so, by the mean value theorem

$$\frac{(2m+1)!}{h^{2m}} \left[\frac{F(x+h) - F(x-h)}{2} - \sum_{i=0}^{m-1} \frac{h^{2i+1}}{(2i+1)!} F_{(2i+1)}^{(s)}(x) \right]$$

$$= (2m+1)\frac{\psi(h)}{h} - \frac{(4m^2-1)}{h^{2m}} \int_0^h t^{2m-2}\psi(t)\, dt$$

$$= (2m+1)\frac{\psi(h)}{h} - \frac{(4m^2-1)}{2m} \frac{\psi(\xi)}{\xi}, \qquad (2.8.19)$$

where $0 < \xi < h$. So, letting $h \to 0$, we have from (2.8.17) and (2.8.19) that

$$\overline{S}_{2m+1}F(x) \le (2m+1)\overline{BS}_{2m}f(x) - \frac{4m^2 - 1}{2m}\underline{BS}_{2m}f(x)$$

and

$$\underline{S}_{2m+1}F(x) \ge (2m+1)\underline{BS}_r f(x) - \frac{4m^2 - 1}{2m}\overline{BS}_{2m}f(x), \qquad (2.8.20)$$

which completes the proof. □

Theorem 2.8.6 *Let f be special Denjoy integrable in some neighbourhood of x and let F be its indefinite integral in that neighbourhood. If $F_{(k)}^{(s)}(x)$ exists, $k \geq 2$, then $SBD_{k-1}f(x)$ exists and $SBD_{k-1}f(x) = F_{(k)}^{(s)}(x)$. More generally, if F is smooth at x of order k, then*

$$\underline{F}_{(k)}^{(s)}(x) \leq \underline{SBD}_{k-1}f(x) \leq \overline{SBD}_{k-1}f(x) \leq \overline{F}_{(k)}^{(s)}(x). \tag{2.8.21}$$

□   We prove the theorem when k is even, the proof when k is odd being similar. Further, we only prove the right-hand inequality in (2.8.20), the left being similar and the inequality implying the first part of the theorem.

First, consider the case of $k = 2$.

We may suppose that $\overline{F}_{(2)}^{(s)}(x) < \infty$ and then choose M such that $\overline{F}_{(2)}^{(s)}(x) < M$. Then there is a $\delta_1 > 0$ such that

$$\frac{F(x+t) + F(x-t) - 2F(x)}{t^2} < M, \text{ for } 0 < t < \delta_1. \tag{2.8.22}$$

Since F is an indefinite integral of f in some neighbourhood of x, then $F' = f$ almost everywhere in that neighbourhood. So, there is a $\delta_2 > 0$ such that for almost $t \in (0, \delta_2)$

$$\left(\frac{F(x+t) + F(x-t) - 2F(x)}{t} \right)' = \frac{f(x+t) - f(x-t)}{t}$$
$$- \frac{F(x+t) + F(x-t) - 2F(x)}{t^2}. \tag{2.8.23}$$

Let $0 < p < q < \min\{\delta_1, \delta_2\}$. Integrating (2.8.22) between p and q and then dividing by q, we have

$$\frac{F(x+q) + F(x-q) - 2F(x)}{q^2} - \frac{F(x+p) + F(x-p) - 2F(x)}{pq}$$
$$= \frac{1}{q} \int_p^q \frac{f(x+t) - f(x-t)}{t} \, dt - \frac{1}{q} \int_p^q \frac{F(x+t) + F(x-t) - 2F(x)}{t^2} \, dt,$$

and so

$$\frac{1}{q} \int_p^q \frac{f(x+t) - f(x-t)}{t} \, dt = \varpi_2(F; x, q) + \frac{1}{q} \int_p^q \varpi_2(F; x, t) \, dt - \frac{p}{q} \varpi_2(F; x, p).$$

Hence, by (2.8.21),

$$\frac{1}{q} \int_p^q \frac{f(x+t) - f(x-t)}{t} \, dt < M + \frac{1}{q} M(q - p) - \frac{p}{q} \varpi_2(F; x, p).$$

Since F is smooth of order 2 at x, $p\varpi_2(F; x, p) \to 0$ as $p \to 0$. So letting $p \to 0$,

$$\limsup_{p \to 0} \frac{1}{q} \int_p^q \frac{f(x+t) - f(x-t)}{t} \, dt \leq 2M.$$

Now, letting $q \to 0$, we get that $\overline{SBD}_1 f(x) \le M$, which, since M was arbitrary, proves the right-hand inequality in this case.

Suppose now that the right-hand inequality holds for $k = 2, 4, \cdots, 2m - 2$. Then we now deduce it holds for $k = 2m$. This will complete the proof by induction.

As before, choose M such that $\overline{F}_{(2m)}^{(s)}(x) < M < \infty$. Then there is $\delta_1 > 0$ such that

$$\varpi_{2m}(F; x, t) < M, \text{ for } 0 < t < \delta_1. \tag{2.8.24}$$

Since F is an indefinite integral of f in some neighbourhood of x, then $F' = f$ almost everywhere in that neighbourhood. So, there is a $\delta_2 > 0$ such that for almost $t \in (0, \delta_2)$

$$\left(\frac{t}{2m}\varpi_{2m}(F; x, t)\right)' = \frac{(2m-1)!}{t^{2m-1}}\left[\frac{f(x+t) - f(x-t)}{2} - \sum_{i=1}^{m-1}\frac{t^{2i-1}}{(2i-1)!}F_{(2i)}^{(s)}(x)\right]$$

$$-\left[\frac{F(x+t) + F(x-t)}{2} - \sum_{i=0}^{m-1}\frac{t^{2i}}{(2i)!}F_{(2i)}^{(s)}(x)\right]\frac{2m-1}{2m}\frac{(2m)!}{t^{2m}}. \tag{2.8.25}$$

Since F is smooth of order $2m$ at x, we have by definition that $F_{(2i)}^{(s)}(x)$ exists finitely with value $SBD_{2i-1}f(x)$ for $i = 1, 2, \cdots, m-1$, and, so, from (2.8.24), that for almost all $t \in (0, \delta_2)$

$$\left(\frac{t}{2m}\varpi_{2m}(F; x, t)\right)' = \overline{\varpi}_{2m-1}(f; x, t) - \frac{2m-1}{2m}\varpi_{2m}(F; x, t). \tag{2.8.26}$$

Let $0 < p < q < \min\{\delta_1, \delta_2\}$. Integrating (2.8.25) between p and q and then multiplying by $2m/q$, we have

$$\varpi_{2m}(F; x, q) - \frac{p}{q}\varpi_{2m}(F; x, p)$$

$$= \frac{2m}{q}\int_p^q \overline{\varpi}_{2m-1}(f; x, t)\, dt - \frac{2m-1}{q}\int_p^q \varpi_{2m}(F; x, t)\, dt,$$

or

$$\frac{2m}{q}\int_p^q \overline{\varpi}_{2m-1}(f; x, t)\, dt$$

$$= \varpi_{2m}(F; x, q) + \frac{2m-1}{q}\int_p^q \varpi_{2m}(F; x, t)\, dt - \frac{p}{q}\varpi_{2m}(F; x, p).$$

Hence, by (2.8.23),

$$\frac{2m}{q}\int_p^q \overline{\varpi}_{2m-1}(f; x, t)\, dt < M + \frac{2m-1}{q}M(q-p) - \frac{p}{q}\varpi_{2m}(F; x, p).$$

Since F is smooth of order $2m$ at x, letting $p \to 0$, we have

$$\frac{2m}{q}\limsup_{p\to 0}\int_p^q \overline{\varpi}_{2m-1}(f; x, t)\, dt \le 2mM.$$

Now, letting $q \to 0$, $2m\overline{SBD}_{2m-1}f(x) \le 2mM$, that is, $\overline{SBD}_{2m-1}f(x) \le M$. Since M is arbitrary, this proves that $\overline{SBD}_{2m-1}f(x) \le \overline{F}_{(2m)}^{(s)}(x)$. □

2.9 Borel and Symmetric Borel Derivatives, $BD_k f$ and $SBD_k f$, and Borel Smoothness of Order k

Theorem 2.9.1 *If $BD_r f(x)$ exists finitely, then $SBD_r f(x)$ exists with the same value. Moreover for small h and $0 < \epsilon < h$*

(i) $\dfrac{1}{2} \displaystyle\int_\epsilon^h \left[\overline{\gamma}_{r+1}(f;x,t) + \overline{\gamma}_{r+1}(f;x,-t) \right] dt = \int_\epsilon^h \overline{\varpi}_{r+1}(f;x,t)\, dt;$

(ii) $\dfrac{1}{2} \displaystyle\int_\epsilon^h t \left[\overline{\gamma}_{r+1}(f;x,t) + \overline{\gamma}_{r+1}(f;x,-t) \right] dt = \int_\epsilon^h t\overline{\varpi}_{r+1}(f;x,t)\, dt;$

(iii) $\dfrac{1}{2} \displaystyle\int_\epsilon^h \left[\overline{\gamma}_{r+1}(f;x,t) - \overline{\gamma}_{r+1}(f;x,-t) \right] dt = \dfrac{1}{r+2} \int_\epsilon^h t\overline{\varpi}_{r+2}(f;x,t)\, dt;$

where

$$\overline{\gamma}_{r+1}(f;x,t) = \frac{(r+1)!}{t^{r+1}} \left[f(x+t) - \sum_{i=0}^r \frac{t^i}{i!} BD_i f(x) \right] \qquad (2.9.1)$$

and $\overline{\varpi}_k(f;x,t)$ is as defined in (1.11.6) of Chapter I.

□ Let $BD_r f(x)$ exist finitely. Then from the definition of $BD_r f(x)$,

$$\frac{1}{h} \int_0^h \frac{f(x+t) - \sum_{i=0}^r \frac{t^i}{i!} BD_i f(x)}{t^r}\, dt = o(1), \quad \text{as } h \to 0. \qquad (2.9.2)$$

Let r be even, $r = 2m$, say. Then from (2.9.2)

$$\frac{1}{h} \int_0^h \frac{\frac{f(x+t) + f(x-t)}{2} - \sum_{i=0}^m \frac{t^{2i}}{(2i)!} BD_{2i} f(x)}{t^{2m}}\, dt$$

$$= \frac{1}{2h} \int_0^h \frac{f(x+t) - \sum_{i=0}^{2m} \frac{t^i}{i!} BD_i f(x)}{t^{2m}}\, dt$$

$$+ \frac{1}{2h} \int_0^h \frac{f(x-t) - \sum_{i=0}^{2m} \frac{(-t)^i}{i!} BD_i f(x)}{(-t)^{2m}}\, dt$$

$$= o(1), \quad \text{as } h \to 0.$$

So, $SBD_{2m}f(x)$ exists and $SBD_{2i}f(x) = BD_{2i}f(x)$ for $i = 0, 1, \cdots, m$. Also, from (2.9.1) and the definition of $\overline{\varpi}_k(f; x, t)$

$$\frac{1}{2}\int_\epsilon^h \left[\overline{\gamma}_{2m+1}(f; x, t) + \overline{\gamma}_{2m+1}(f; x, -t)\right] dt$$

$$= \frac{1}{2}\int_\epsilon^h \frac{(2m+1)!}{t^{2m+1}}\left[f(x+t) - \sum_{i=0}^{2m} \frac{t^i}{i!}BD_if(x) - f(x-t)\right.$$

$$\left. + \sum_{i=0}^{2m} \frac{(-t)^i}{i!}BD_if(x)\right] dt$$

$$= \frac{1}{2}\int_\epsilon^h \frac{(2m+1)!}{t^{2m+1}}\left[f(x+t) - f(x-t) - \sum_{i=0}^{2m} \frac{t^i\left(1-(-1)^i\right)}{i!}BD_if(x)\right] dt$$

$$= \int_\epsilon^h \frac{(2m+1)!}{t^{2m+1}}\left[\frac{f(x+t) - f(x-t)}{2} - \sum_{i=0}^{m-1} \frac{t^{2i+1}}{(2i+1)!}BD_{2i+1}f(x)\right] dt$$

$$= \int_\epsilon^h \overline{\varpi}_{2m+1}(f; x, t)\, dt.$$

Similarly,

$$\frac{1}{2}\int_\epsilon^h t\left[\overline{\gamma}_{2m+1}(f; x, t) + \overline{\gamma}_{2m+1}(f; x, -t)\right] dt = \int_\epsilon^h t\overline{\varpi}(f; x, t)\, dt.$$

Also

$$\frac{1}{2}\int_\epsilon^h \left[\overline{\gamma}_{2m+1}(f; x, t) - \overline{\gamma}_{2m+1}(f; x, -t)\right] dt$$

$$= \frac{1}{2}\int_\epsilon^h \frac{(2m+1)!}{t^{2m+1}}\left[f(x+t) - \sum_{i=0}^{2m} \frac{t^i}{i!}BD_if(x) + f(x-t)\right.$$

$$\left. - \sum_{i=0}^{2m} \frac{(-t)^i}{i!}BD_if(x)\right] dt$$

$$= \frac{1}{2}\int_\epsilon^h \frac{(2m+1)!}{t^{2m+1}}\left[f(x+t) + f(x-t) - \sum_{i=0}^{2m} \frac{t^i\left(1+(-1)^i\right)}{i!}BD_if(x)\right] dt$$

$$= \int_\epsilon^h \frac{(2m+1)!}{t^{2m+1}}\left[\frac{f(x+t) + f(x-t)}{2} - \sum_{i=0}^{m} \frac{t^{2i}}{(2i)!}BD_{2i}f(x)\right] dt$$

$$= \int_\epsilon^h \frac{t}{2m+2}\overline{\varpi}_{2m+2}(f; x, t)\, dt.$$

This proves (i), (ii) and (iii) when r is even.

The proof when r is odd is similar. $\square$

Corollary 2.9.2 *If $BD_r f(x)$ exists, possibly infinite, then $SBD_r f(x)$ exists with the same value. If f is Borel smooth of order $r+1$ at x, then $\overline{BD}_r^+ f(x) = \overline{BD}_r^- f(x)$ and $\underline{BD}_r^+ f(x) = \underline{BD}_r^- f(x)$.*

□   The proof of the first part follows from Theorem 2.9.1 (i) and that of the second part follows from Theorem 2.9.1 (iii).   □

Theorem 2.9.3 *If f is Borel smooth of orders $r + 1$ and $r + 2$ at x, then $BD_r f(x)$ exists finitely. If $BD_r f(x)$ exists finitely, then f is Borel smooth of order $r + 1$ at x.*

□   We prove the theorem for even r; the proof is similar when r is odd. Let $r = 2m$. Then, since f is Borel smooth of order $2m + 1$ at x, $SBD_{2m-1}f(x)$ exists finitely and as $h \to 0+$,

$$\int_0^h \frac{(2m+1)!}{t^{2m}}\left[\frac{f(x+t) - f(x-t)}{2} - \sum_{i=0}^{m-1} \frac{t^{2i+1}}{(2i+1)!}SBD_{2i+1}f(x)\right] dt = o(h).$$

$$(2.9.3)$$

Again, since f is Borel smooth of order $2m+2$ at x, $SBD_{2m}f(x)$ exists finitely and as $h \to 0+$,

$$\int_0^h \frac{(2m+2)!}{t^{2m+1}}\left[\frac{f(x+t) + f(x-t)}{2} - \sum_{i=0}^{m} \frac{t^{2i}}{(2i)!}SBD_{2i}f(x)\right] dt = o(h),$$

and, so, by Lemma 2.8.4 as $h \to 0+$,

$$\int_0^h \frac{(2m+2)!}{t^{2m}}\left[\frac{f(x+t) + f(x-t)}{2} - \sum_{i=0}^{m} \frac{t^{2i}}{(2i)!}SBD_{2i}f(x)\right] dt = o(h^2).$$

$$(2.9.4)$$

From (2.9.3) and (2.9.4), we have

$$\int_0^h \frac{1}{t^{2m}}\left[f(x+t) - \sum_{i=0}^{2m} \frac{t^i}{i!}SBD_i f(x)\right] dt = o(h),$$

which shows that $BD_{2m}f(x)$ exists with value $SBD_{2m}f(x)$.

Now, suppose that $BD_r f(x)$ exists finitely. Then $\int_0^h [\overline{\gamma}_r(f(;x,t) - BD_r f(x)]\, dt = o(h)$ as $h \to 0$. Hence, since $t\overline{\gamma}_{r+1}(f(;x,t) = (r + 1)[\overline{\gamma}_r(f(;x,t) - BD_r f(x)]$ we have that $\int_0^h t\overline{\gamma}_{r+1}(f(;x,t)\, dt = o(h)$ as $h \to 0+$, and so, applying Theorem 2.9.1(ii), we have $\int_0^h t\overline{\omega}_{r+1}(f(;x,t)\, dt = o(h)$ as $h \to 0+$, showing that f is Borel smooth of order $r + 1$ at x.   □

2.10   Peano and L^p-Derivatives, $f_{(k)}$ and $f_{(k),p}$

Theorem 2.10.1 *Let $f \in \mathcal{L}^p$, $1 \leq p < \infty$, in some neighbourhood of x. If $f_{(r)}(x)$ exists finitely, then $f_{(r),p}(x)$ exists with the same value, but the converse is false.*

☐   Since $f_{(r)}(x)$ exists finitely, we have from (1.4.2) of Chapter I

$$f(x+t) - \sum_{i=0}^{r} \frac{t^i}{i!} f_{(i)}(x) = o(t^r) \text{ as } t \to 0.$$

Let $\epsilon > 0$ be arbitrary. Then there is a $\delta > 0$ such that

$$\left| \frac{f(x+t) - \sum_{i=0}^{r} \frac{t^i}{i!} f_{(i)}(x)}{t^r} \right| < \epsilon \text{ for } 0 < |t| < \delta. \tag{2.10.1}$$

Let $0 < h < \delta$. Then from (2.10.1)

$$\left| f(x+t) - \sum_{i=0}^{r} \frac{t^i}{i!} f_{(i)}(x) \right|^p < (\epsilon t^r)^p \text{ for } 0 < t < h,$$

and

$$\left| f(x-\xi) - \sum_{i=0}^{r} \frac{(-\xi)^i}{i!} f_{(i)}(x) \right|^p < (\epsilon \xi^r)^p \text{ for } 0 < \xi < h.$$

Hence,

$$\left[\frac{1}{h} \int_0^h \left| f(x+t) - \sum_{i=0}^{r} \frac{t^i}{i!} f_{(i)}(x) \right|^p dt \right]^{1/p} \leq \left[\epsilon^p \frac{h^{rp}}{rp+1} \right]^{1/p} = \epsilon \frac{h^r}{(rp+1)^{1/p}}, \tag{2.10.2}$$

and

$$\left[\frac{1}{h} \int_0^h \left| f(x-\xi) - \sum_{i=0}^{r} \frac{(-\xi)^i}{i!} f_{(i)}(x) \right|^p d\xi \right]^{1/p} \leq \epsilon \frac{h^r}{(rp+1)^{1/p}}. \tag{2.10.3}$$

Changing variables in (2.10.3) we have

$$\left[\frac{1}{-h} \int_0^{-h} \left| f(x+t) - \sum_{i=0}^{r} \frac{t^i}{i!} f_{(i)}(x) \right|^p dt \right]^{1/p} \leq \epsilon \frac{h^r}{(rp+1)^{1/p}}. \tag{2.10.4}$$

From (2.10.2) and (2.10.4)

$$\left[\frac{1}{h} \int_0^h \left| f(x+t) - \sum_{i=0}^{r} \frac{t^i}{i!} f_{(i)}(x) \right|^p dt \right]^{1/p} = o(h^r), \text{ as } h \to 0.$$

Hence, $f_{(r),p}(x)$ exists and $f_{(r),p}(x) = f_{(r)}(x)$.

For the converse, consider the function

$$f(x) = \begin{cases} x^r & \text{if } x = 0 \text{ or } x \text{ is irrational,} \\ 1 & \text{if } x \text{ is rational and } x \neq 0 \, . \end{cases}$$

Then $\dfrac{1}{h}\displaystyle\int_0^h \left| f(t) - t^r \right| \, dt = 0$ so $f_{(r),p}(0)$ exists with value $r!$, while $f_{(i),p}(0) = 0$, $i = 1, 2, \cdots, r-1$. However, $f_{(1)}(0)$ does not exist and so $f_{(r)}(0)$ cannot exist. $\qquad\square$

Theorem 2.10.2 *Let $f \in \mathcal{L}^p$, $1 \leq p < \infty$, in some neighbourhood of x and let F be an indefinite integral of f in that neighbourhood. If $f_{(r),p}(x)$ exists finitely, then $F_{(r+1)}(x)$ exists with value $f_{(r),p}(x)$.*

$\square\qquad$ By Hölder's inequality,

$$\int_0^h \left| f(x+t) - \sum_{i=0}^r \frac{t^i}{i!} f_{(i),p}(x) \right| dt \leq \left[\int_0^h \left| f(x+t) - \sum_{i=0}^r \frac{t^i}{i!} f_{(i),p}(x) \right|^p dt \right]^{1/p} h^{1-1/p},$$

and so,

$$\frac{1}{h}\int_0^h \left| f(x+t) - \sum_{i=0}^r \frac{t^i}{i!} f_{(i),p}(x) \right| dt \leq \left[\frac{1}{h}\int_0^h \left| f(x+t) - \sum_{i=0}^r \frac{t^i}{i!} f_{(i),p}(x) \right|^p dt \right]^{1/p}.$$

When $h < 0$, put $k = -h$ to obtain

$$\frac{1}{k}\int_0^k \left| f(x+t) - \sum_{i=0}^r \frac{t^i}{i!} f_{(i),p}(x) \right| dt \leq \left[\frac{1}{k}\int_0^k \left| f(x+t) - \sum_{i=0}^r \frac{t^i}{i!} f_{(i),p}(x) \right|^p dt \right]^{1/p},$$

and, so, for any small $|h| > 0$

$$\left| \frac{1}{h}\int_0^h \left[f(x+t) - \sum_{i=0}^r \frac{t^i}{i!} f_{(i),p}(x) \right] dt \right| \leq \frac{1}{h}\int_0^h \left| f(x+t) - \sum_{i=0}^r \frac{t^i}{i!} f_{(i),p}(x) \right| dt$$

$$\leq \left[\frac{1}{h}\int_0^h \left| f(x+t) - \sum_{i=0}^r \frac{t^i}{i!} f_{(i),p}(x) \right|^p dt \right]^{1/p} = o(h^r), \text{ as } h \to 0.$$

Hence,

$$\int_0^h \left[f(x+t) - \sum_{i=0}^r \frac{t^i}{i!} f_{(i),p}(x) \right] dt = o(h^{r+1}), \text{ as } h \to 0,$$

which gives

$$F(x+h) - F(x) - \sum_{i=0}^r \frac{h^{i+1}}{(i+1)!} f_{(i),p}(x) = o(h^{r+1}), \text{ as } h \to 0,$$

which shows that $F_{(r+1)}(x)$ exists and equals $f_{(r),p}(x)$. $\qquad\square$

Theorem 2.10.3 Let $f \in \mathcal{L}^p$, $1 \le p < \infty$, in some neighbourhood of x and let F be an indefinite integral of f in that neighbourhood. If f is L^p bounded of order r at x, then $F_{(r)}(x)$ exists finitely and both $\overline{F}_{(r+1)}(x)$ and $\underline{F}_{(r+1)}(x)$ are finite.

□   By Corollary 1.12.5 of Chapter I, $f_{(r-1),p}(x)$ exists and

$$\left[\frac{1}{h}\int_0^h \left| f(x+t) - \sum_{i=0}^{r-1} \frac{t^i}{i!} f_{(i),p}(x) \right|^p dt \right]^{1/p} = O(h^r), \text{ as } h \to 0.$$

So, proceeding as in Theorem 2.10.2, we have

$$\int_0^h \left[f(x+t) - \sum_{i=0}^{r-1} \frac{t^i}{i!} f_{(i),p}(x) \right] dt = O(h^{r+1}), \text{ as } h \to 0,$$

which gives

$$F(x+h) - F(x) - \sum_{i=0}^{r-1} \frac{h^{i+1}}{(i+1)!} f_{(i),p}(x) = O(h^{r+1}), \text{ as } h \to 0.$$

So, F is Peano bounded of order $r+1$. The result then follows by Theorem 1.4.4 of Chapter I.   □

Theorem 2.10.4 Let $f \in \mathcal{L}^p$, $1 \le p < \infty$, in some neighbourhood of x. If f is Peano bounded of order r at x, then f is L^p bounded of order r at x. The converse is not true.

□   Since f is Peano bounded of order r at x by Theorem 1.4.4 of Chapter I, $f_{(r-1)}(x)$ exits finitely and

$$f(x+t) - \sum_{i=0}^{r-1} \frac{t^i}{i!} f_{(i)}(x) = O(t^r), \text{ as } t \to 0.$$

So, there is an $M > 0$ and a $\delta > 0$ such that

$$\left| \frac{1}{t^r}\left[f(x+t) - \sum_{i=0}^{r-1} \frac{t^i}{i!} f_{(i)}(x) \right] \right| < M, \text{ for } 0 < |t| < \delta. \tag{2.10.5}$$

This relation is analogous to (2.10.1). Proceeding as in the proof of Theorem 2.10.1, we get from (2.10.6)

$$\left[\frac{1}{h}\int_0^h \left| f(x+t) - \sum_{i=0}^{r-1} \frac{t^i}{i!} f_{(i)}(x) \right|^p dt \right]^{1/p} = O(h^r), \text{ as } h \to 0,$$

which shows that f is L^p-bounded of order r at x.

The converse part is similar to that of Theorem 2.10.1.   □

2.11 L^p- and Symmetric L^p-Derivatives, $f_{(k),p}$ and $f^{(s)}_{(k),p}$

Theorem 2.11.1 *If* $f_{(r),p}(x)$ *exists, then* $f^{(s)}_{(r),p}(x)$ *exists with the same value, but the converse is false.*

☐   Since $f_{(r),p}(x)$ exists, then

$$\left[\frac{1}{h}\int_0^h \left|f(x+t) - \sum_{i=0}^r \frac{t^i}{i!}f_{(i),p}(x)\right|^p dt\right]^{1/p} = o(h^r), \text{ as } h \to 0. \qquad (2.11.1)$$

Let r be even, $r = 2m$, say. Then

$$\frac{f(x+t) + f(x-t)}{2} - \sum_{i=0}^m \frac{t^{2i}}{(2i)!}f_{(2i),p}(x)$$

$$= \frac{f(x+t) - \sum_{i=0}^{2m}\frac{t^i}{i!}f_{(i),p}(x)}{2} + \frac{f(x-t) - \sum_{i=0}^{2m}\frac{(-t)^i}{i!}f_{(i),p}(x)}{2}.$$

Hence, by Minkowski's inequality and by (2.11.1),

$$\left[\frac{1}{h}\int_0^h\left|\frac{f(x+t) + f(x-t)}{2} - \sum_{i=0}^m \frac{t^{2i}}{(2i)!}f_{(2i),p}(x)\right|^p dt\right]^{1/p}$$

$$\leq \left[\frac{1}{h}\int_0^h\left|\frac{f(x+t) - \sum_{i=0}^{2m}\frac{t^i}{i!}f_{(i),p}(x)}{2}\right|^p dt\right]^{1/p}$$

$$+ \left[\frac{1}{h}\int_0^h\left|\frac{f(x-t) - \sum_{i=0}^{2m}\frac{(-t)^i}{i!}f_{(i),p}(x)}{2}\right|^p dt\right]^{1/p} = o(h^r), \text{ as } h \to 0,$$

and so, $f^{(s)}_{(2m),p}(x)$ exists and is equal to $f_{(2m),p}(x)$.

If r is odd, $r = 2m + 1$, say, then

$$\frac{f(x+t) - f(x-t)}{2} - \sum_{i=0}^m \frac{t^{2i+1}}{(2i+1)!}f_{(2i+1),p}(x)$$

$$= \frac{f(x+t) - \sum_{i=0}^{2m+1}\frac{t^i}{i!}f_{(i),p}(x)}{2} - \frac{f(x-t) - \sum_{i=0}^{2m+1}\frac{(-t)^i}{i!}f_{(i),p}(x)}{2},$$

and hence, as above,

$$\left[\frac{1}{h}\int_0^h \left|\frac{f(x+t)-f(x-t)}{2} - \sum_{i=0}^{m}\frac{t^{2i+1}}{(2i+1)!}f_{(2i+1),p}(x)\right|^p dt\right]^{1/p}$$

$$\leq \left[\frac{1}{h}\int_0^h \left|\frac{f(x+t)-\sum_{i=0}^{2m+1}\frac{t^i}{i!}f_{(i),p}(x)}{2}\right|^p dt\right]^{1/p}$$

$$+\left[\frac{1}{h}\int_0^h \left|\frac{f(x-t)-\sum_{i=0}^{2m+1}\frac{(-t)^i}{i!}f_{(i),p}(x)}{2}\right|^p dt\right]^{1/p} = o(h^r), \text{ as } h \to 0,$$

and so $f_{(2m+1),p}^{(s)}(x)$ exists and is equal to $f_{(2m+1),p}(x)$.

Regarding the converse, let

$$f(x) = \begin{cases} 1, & \text{if } x \neq 0, \\ 0, & \text{if } x = 0 . \end{cases}$$

If $f_{(1),1}(0)$ exist, with value λ say, then

$$\frac{1}{h}\int_0^h |f(t) - t\lambda| \, dt = o(h), \text{ as } h \to 0. \tag{2.11.2}$$

Let $h > 0$ be so small that $1 - \lambda t > \frac{1}{2}$ for $0 < t < h$. Then, by (2.11.2),

$$\frac{1}{2} < 1 - \lambda\frac{h}{2} = \frac{1}{h}\int_0^h (1-\lambda t)\, dt = \frac{1}{h}\int_0^h |f(t) - t\lambda|\, dt = o(h), \text{ as } h \to 0.$$

This contradiction shows that $f_{(1),1}(0)$ does not exist and, therefore, by Theorem 2.13.3 of Chapter I, $f_{(1),p}(0)$ cannot exist for any $p \geq 1$ and, so, $f_{(r),p}(0)$ cannot exist for any $r \geq 1$. However, for all h

$$\frac{1}{h}\int_0^h \left|\frac{f(t)-f(-t)}{2}\right|^p dt = 0$$

and, hence, $f_{(2m+1),p}^{(s)}(0)$ exists, with value 0, for all $m = 0, 1, \ldots$. □

Theorem 2.11.2 *If f is L^p-smooth of orders $r+1$ and $r+2$ at x, then $f_{(r),p}(x)$ exists. If $f_{(r),p}(x)$ exists, then f is L^p-smooth of order $r+1$ at x.*

□   We consider the case of r even, $r = 2m$, say. The case of odd r can be treated similarly.

Since f is L^p-smooth of order $2m+1$ at x, $f_{(2m-1),p}^{(s)}(x)$ exists and as $h \to 0$,

$$\left[\frac{1}{h}\int_0^h \left|\frac{f(x+t)-f(x-t)}{2} - \sum_{i=0}^{m-1}\frac{t^{2i+1}}{(2i+1)!}f_{(2i+1),p}^{(s)}(x)\right|^p dt\right]^{1/p} = o(h^{2m})$$

$$\tag{2.11.3}$$

(see the relation (2.14.7) of Chapter I). Similarly, the smoothness of order $2m + 2$ at x gives, as $h \to 0$,

$$\left[\frac{1}{h}\int_0^h \left|\frac{f(x+t)+f(x-t)}{2} - \sum_{i=0}^m \frac{t^{2i}}{(2i)!}f_{(2i),p}^{(s)}(x)\right|^p dt\right]^{1/p} = o(h^{2m+1}).$$

(2.11.4)

So, by Minkowski's inequality, we have from (2.11.3) and (2.11.4) that

$$\left[\frac{1}{h}\int_0^h \left|f(x+t) - \sum_{i=0}^m \frac{t^{2i}}{(2i)!}f_{(2i),p}^{(s)}(x) - \sum_{i=0}^{m-1} \frac{t^{2i+1}}{(2i+1)!}f_{(2i+1),p}^{(s)}(x)\right|^p dt\right]^{1/p}$$

$$\leq \left[\frac{1}{h}\int_0^h \left|\frac{f(x+t)+f(x-t)}{2} - \sum_{i=0}^m \frac{t^{2i}}{(2i)!}f_{(2i),p}^{(s)}(x)\right|^p dt\right]^{1/p}$$

$$+ \left[\frac{1}{h}\int_0^h \left|\frac{f(x+t)-f(x-t)}{2} - \sum_{i=0}^{m-1} \frac{t^{2i+1}}{(2i+1)!}f_{(2i+1),p}^{(s)}(x)\right|^p dt\right]^{1/p}$$

$$= o(h^{2m+1}) + o(h^{2m}) = o(h^{2m}), \text{ as } h \to 0.$$

This shows that $f_{(2m),p}(x)$ exists and that $f_{(i),p}(x) = f_{(i),p}^{(s)}(x), 1 = 0, 1, \cdots, 2m$.

Next, suppose that $f_{(2m),p}(x)$ exists. Then

$$\left[\frac{1}{h}\int_0^h \left|f(x+t) - \sum_{i=0}^{2m} \frac{t^i}{i!}f_{(i),p}(x)\right|^p dt\right]^{1/p} = o(h^{2m}), \text{ as } h \to 0. \quad (2.11.5)$$

Then applying Minkowski's inequality and (2.11.5),

$$\left[\frac{1}{h}\int_0^h \left|\frac{f(x+t)-f(x-t)}{2} - \sum_{i=0}^{m-1} \frac{t^{2i+1}}{(2i+1)!}f_{(2i+1),p}(x)\right|^p dt\right]^{1/p}$$

$$= \frac{1}{2}\left[\frac{1}{h}\int_0^h \left|f(x+t) - \sum_{i=0}^{2m} \frac{t^i}{i!}f_{(i),p}(x) - f(x-t) + \sum_{i=0}^{2m} \frac{(-t)^i}{i!}f_{(i),p}(x)\right|^p dt\right]^{1/p}$$

$$\leq \frac{1}{2}\left(\left[\frac{1}{h}\int_0^h \left|f(x+t) - \sum_{i=0}^{2m} \frac{t^i}{i!}f_{(i),p}(x)\right|^p dt\right]^{1/p}\right.$$

$$\left. + \left[\frac{1}{h}\int_0^h \left|f(x-t) - \sum_{i=0}^{2m} \frac{(-t)^i}{i!}f_{(i),p}(x)\right|^p dt\right]^{1/p}\right)$$

$$= o(h^{2m}), \text{ as } h \to 0.$$

This shows that f is L^p-smooth of order $2m + 1$ at x. $\qquad \square$

Theorem 2.11.3 If f is L^p-bounded of order r at x, then f is symmetric L^p-bounded of order r at x, but not conversely.

□   Since f is L^p-bounded of order r at x, by Corollary 1.12.5 of Chapter I, $f_{(r-1),p}(x)$ exists and

$$\left[\frac{1}{h}\int_0^h \left|f(x+t) - \sum_{i=0}^{r-1}\frac{t^i}{i!}f_{(i),p}(x)\right|^p dt\right]^{1/p} = O(h^r), \text{ as } h \to 0. \qquad (2.11.6)$$

Suppose that r is even, $r = 2m$, say. Then, by Minkowski's inequality and (2.11.6),

$$\left[\frac{1}{h}\int_0^h \left|\frac{f(x+t)+f(x-t)}{2} - \sum_{i=0}^{m-1}\frac{t^{2i}}{(2i)!}f_{(2i),p}(x)\right|^p dt\right]^{1/p}$$

$$= \frac{1}{2}\left[\frac{1}{h}\int_0^h\left|f(x+t) - \sum_{i=0}^{2m-1}\frac{t^i}{i!}f^{(s)}_{(i),p}(x) + f(x-t) - \sum_{i=0}^{2m-1}\frac{(-t)^i}{i!}f^{(s)}_{(i),p}(x)\right|^p dt\right]^{1/p}$$

$$\leq \frac{1}{2}\left(\left[\frac{1}{h}\int_0^h\left|f(x+t) - \sum_{i=0}^{2m-1}\frac{t^i}{i!}f^{(s)}_{(i),p}(x)\right|^p dt\right]^{1/p}\right.$$

$$\left. + \left[\frac{1}{h}\int_0^h\left|f(x-t) - \sum_{i=0}^{2m-1}\frac{(-t)^i}{i!}f^{(s)}_{(i),p}(x)\right|^p dt\right]^{1/p}\right)$$

$$= O(h^{2m}), \text{ as } h \to 0.$$

This proves the result for even r and the case of odd r is similar. The converse is clear.   □

2.12   Symmetric de la Vallée Poussin and Symmetric L^p-Derivatives, $f^{(s)}_{(k)}$ and $f^{(s)}_{(k),p}$

Theorem 2.12.1 *Let $f \in L^p$, $1 \leq p < \infty$, in some neighbourhood of x. If $f^{(s)}_{(r)}(x)$ exists finitely, then $f^{(s)}_{(r),p}(x)$ exists and is equal to $f^{(s)}_{(r)}(x)$. The converse is not true.*

□   Since $f^{(s)}_{(r)}(x)$ exists finitely, we have from relation (1.6.1) of Chapter I,

$$\frac{f(x+t)+(-1)^r f(x-t)}{2} - P(t) = o(t^r) \text{ as } t \to 0,$$

where $P(t)$ is the polynomial

$$P(t) = \frac{t^r}{r!}f^{(s)}_{(r)}(x) + \frac{t^{r-2}}{(r-2)!}f^{(s)}_{(r-2)}(x) + \cdots \qquad (2.12.1)$$

(see the relation (1.6.6) of Chapter I).

Let $\epsilon > 0$ be arbitrary. Then there is a $\delta > 0$ such that

$$\left| \frac{1}{t^r} \left[\frac{f(x+t) + (-1)^r f(x-t)}{2} - P(t) \right] \right| < \epsilon \text{ for } 0 < |t| < \delta.$$

Let $0 < h < \delta$. Then $\left| \frac{1}{2} (f(x+t) + (-1)^r f(x-t)) - P(t) \right|^p < (\epsilon t^r)^p$ for $0 < t < h$. Hence,

$$\left[\frac{1}{h} \int_0^h \left| \frac{1}{2} (f(x+t) + (-1)^r f(x-t)) - P(t) \right|^p dt \right]^{1/p} \leq \left(\frac{\epsilon^p h^{rp}}{rp+1} \right)^{1/p}. \quad (2.12.2)$$

Similarly, if $-\delta < h < 0$,

$$\left[\frac{1}{h} \int_0^h \left| \frac{1}{2} (f(x+t) + (-1)^r f(x-t)) - P(t) \right|^p dt \right]^{1/p} \leq \left(\frac{\epsilon^p (-h)^{rp}}{rp+1} \right)^{1/p}. \quad (2.12.3)$$

From (2.12.2) and (2.12.3),

$$\left[\frac{1}{h} \int_0^h \left| \frac{1}{2} (f(x+t) + (-1)^r f(x-t)) - P(t) \right|^p dt \right]^{1/p} = o(h^r), \text{ as } h \to 0.$$

Hence, $f_{(r),p}^{(s)}(x)$ exists, and from (2.12.1) and from the definition of $f_{(r),p}^{(s)}(x)$, it follows that $f_{(r),p}^{(s)}(x) = f_{(r)}^{(s)}(x)$.

For the converse, consider

$$f(x) = \begin{cases} 1, & \text{if } x > 0 \text{ and rational}, \\ 0, & \text{otherwise}. \end{cases}$$

Then $f_{(1)}^{(s)}(0)$ and $f_{(2)}^{(s)}(0)$ do not exist and, so, $f_{(r)}^{(s)}(0)$ cannot exist for any r, odd or even, but $f_{(r),p}^{(s)}(0)$ exists with $f_{(r),p}^{(s)}(0) = 0$ for any r. $\qquad \square$

Theorem 2.12.2 *Let $f \in \mathcal{L}^p$, $1 \leq p < \infty$, in some neighbourhood of x and let F be an indefinite integral of f in that neighbourhood. If $f_{(r),p}^{(s)}(x)$ exists, then $F_{(r+1)}^{(s)}(x)$ exists with value $f_{(r),p}^{(s)}(x)$.*

$\square$ By Hölder's inequality, we have, when $h > 0$, that

$$\int_0^h \left| \frac{1}{2} (f(x+t) + (-1)^r f(x-t)) - P(t) \right| dt$$

$$\leq \left[\int_0^h \left| \frac{1}{2} (f(x+t) + (-1)^r f(x-t)) - P(t) \right|^p dt \right]^{1/p} h^{1-1/p},$$

where $P(t)$ is given by (see relation (1.13.6) of Chapter I)

$$P(t) = \begin{cases} \sum_{i=0}^{r/2} \frac{t^{2i}}{(2i)!} f_{(2i),p}^{(s)}(x), & \text{if } r \text{ is even}, \\ \sum_{i=0}^{(r-1)/2} \frac{t^{2i+1}}{(2i+1)!} f_{(2i+1),p}^{(s)}(x), & \text{if } r \text{ is odd}, \end{cases}$$

which can be written

$$P(t) = \frac{t^r}{r!} f^{(s)}_{(r),p}(x) + \frac{t^{r-2}}{(r-2)!} f^{(s)}_{(r-2),p}(x) + \cdots . \tag{2.12.4}$$

So,

$$\frac{1}{h} \int_0^h \left| \frac{1}{2}(f(x+t) + (-1)^r f(x-t)) - P(t) \right| dt$$

$$\leq \left[\frac{1}{h} \int_0^h \left| \frac{1}{2}(f(x+t) + (-1)^r f(x-t)) - P(t) \right|^p dt \right]^{1/p}$$

$$= o(h^r), \text{ as } h \to 0+.$$

A similar result holds when $h \to 0-$ and, so,

$$\int_0^h \left(\frac{1}{2}(f(x+t) + (-1)^r f(x-t)) - P(t) \right) dt = o(h^{r+1}), \quad \text{as } h \to 0.$$

So, using (2.12.4), we have

$$\frac{1}{2}\left(F(x+h) + (-1)^{r+1} F(x-h)\right) - \frac{h^{r+1}}{(r+1)!} f^{(s)}_{(r),p}(x) - \cdots = o(h^{r+1}), \quad \text{as } h \to 0.$$

Hence, $F^{(s)}_{(r+1)}(x)$ exists and $F^{(s)}_{(r+1)}(x) = f^{(s)}_{(r),p}(x)$. $\square$

Theorem 2.12.3 *Let $f \in \mathcal{L}^p$, $1 \leq p < \infty$, in some neighbourhood of x and let F be an indefinite integral of f in that neighbourhood. If f is symmetric L^p-bounded of order r at x, then $F^{(s)}_{(r-1)}(x)$ exists and $\overline{F}^{(s)}_{(r+1)}(x)$ and $\underline{F}^{(s)}_{(r+1)}(x)$ are finite.*

$\square$ By Theorem 1.13.3 of Chapter I, $f^{(s)}_{(r-2),p}(x)$ exists and so, by Theorem 2.12.2, $F^{(s)}_{(r-1)}(x)$ exists and further from relation (1.13.4) of Chapter I,

$$\left[\frac{1}{h} \int_0^h \left| \frac{f(x+t) + (-1)^r f(x-t)}{2} \right. \right.$$

$$\left. \left. - \frac{t^{r-2}}{(r-2)!} f^{(s)}_{(r-2),p}(x) - \frac{t^{r-4}}{(r-4)!} f^{(s)}_{(r-4),p}(x) - \cdots \right|^p dt \right]^{1/p}$$

$$= O(h^r), \text{ as } h \to 0.$$

Hence, as in Theorem 2.12.2,

$$\int_0^h \left[\frac{f(x+t) + (-1)^r f(x-t)}{2} - \frac{t^{r-2}}{(r-2)!} f^{(s)}_{(r-2),p}(x) \right.$$

$$\left. - \frac{t^{r-4}}{(r-4)!} f^{(s)}_{(r-4),p}(x) - \cdots \right] dt$$

$$= O(h^{r+1}), \text{ as } h \to 0,$$

and so

$$\frac{1}{2}\big(F(x+h)+(-1)^{r+1}F(x-h)\big)-\frac{h^{r-1}}{(r-1)!}f^{(s)}_{(r-2),p}(x)-\cdots = O(h^{r+1}), \quad \text{as } h \to 0,$$

which shows that $F^{(s)}_{(r-1)}(x)$ exists and $\overline{F}^{(s)}_{(r+1)}(x)$ and $\underline{F}^{(s)}_{(r+1)}(x)$ are finite. $\square$

Theorem 2.12.4 *Let* $f \in L^p$, $1 \le p < \infty$, *in some neighbourhood of* x. *If* f *is smooth of order* r *at* x, *then* f *is* L^p-*smooth of order* r *at* x, *but not conversely.*

$\square$ Let f be smooth of order r at x. Then $f^{(s)}_{(r-2)}(x)$ exists finitely and $h\varpi_r(f;x,h) = o(1)$ as $h \to 0$, where $\varpi_r(f;x,h)$ is defined in relation (1.6.5) of Chapter I. Hence,

$$\frac{1}{2}\big(f(x+t)+(-1)^r f(x-t)\big) - P(t) = o(h^{r-1}) \quad \text{as } h \to 0,$$

where $P(t)$ is given by

$$P(t) = \frac{t^{r-2}}{(r-2)!}f^{(s)}_{(r-2)}(x) + \frac{t^{r-4}}{(r-4)!}f^{(s)}_{(r-4)}(x) + \cdots$$

(see relation (1.6.5) of Chapter I). So, proceeding as in Theorem 2.12.1, we have

$$\left[\frac{1}{h}\int_0^h \left|\frac{1}{2}\big(f(x+t)+(-1)^r f(x-t)\big) - P(t)\right|^p dt\right]^{1/p} = o(h^{r-1}), \quad \text{as } h \to 0.$$

Hence, f is L^p-smooth of order r at x.
    For the converse, the function in Theorem 2.12.1 will suffice. $\square$

Theorem 2.12.5 *Let* $f \in L^p$, $1 \le p < \infty$, *in some neighbourhood of* x. *If* $f^{(s)}_{(r)}(x)$ *exists finitely and if* $\overline{f}^{(s)}_{(r+2)}(x)$ *and* $\underline{f}^{(s)}_{(r+2)}(x)$ *are finite, then* f *is symmetric* L^p-*bounded of order* $r+2$ *at* x.

$\square$ Since $\overline{f}^{(s)}_{(r+2)}(x)$ and $\underline{f}^{(s)}_{(r+2)}(x)$ are finite, we have

$$\frac{1}{2}\big(f(x+t)+(-1)^{r+2}f(x-t)\big) - P(t) = O(t^{r+2}), \quad \text{as } t \to 0,$$

where $P(t)$ is defined in (2.12.1). So, there are $M > 0$ and $\delta > 0$ such that

$$\left|\frac{1}{t^{r+2}}\left[\frac{f(x+t)+(-1)^{r+2}f(x-t)}{2} - P(t)\right]\right| < M, \quad \text{for } 0 < |t| < \delta.$$

Proceeding as in the proof of Theorem 2.12.1, replacing ϵ and r by M and $r+2$, we have for $0 < |h| < \delta$

$$\left[\frac{1}{h}\int_0^h \left|\frac{1}{2}\big(f(x+t)+(-1)^{r+2}f(x-t)\big) - P(t)\right|^p dt\right]^{1/p} \le \left(\frac{M^p h^{(r+2)p}}{(r+2)p+1}\right)^{1/p},$$

and, hence,

$$\left[\frac{1}{h}\int_0^h\left|\frac{1}{2}(f(x+t)+(-1)^{r+2}f(x-t))-P(t)\right|^p dt\right]^{1/p}=O(h^{r+2}),, \text{ as } h\to 0.$$

So, f is symmetric L^p-bounded of order $r+2$ at x. $\qquad\square$

Corollary 2.12.6 *If $f \in \mathcal{L}^p$, $1 \le p < \infty$, in some neighbourhood of x and if f is d.l.V.P bounded of order r at x, then f is symmetric L^p-bounded of order r at x.*

2.13 Borel and L^p-Derivatives, $BD_{(k)}f$ and $f_{(k),p}$

Theorem 2.13.1 *If $f_{(r),p}(x)$ exists, so does $BD_r f(x)$ with the same value; the converse is not true.*

$\square\qquad$ Let $f_{(r),p}(x)$ exist. Then

$$\left[\frac{1}{h}\int_0^h\left|f(x+t)-\sum_{i=0}^r\frac{t^i}{i!}f_{(i),p}(x)\right|^p dt\right]^{1/p}=o(h^r), \text{ as } h\to 0.$$

So, applying Hölder's inequality, we get as in (2.10.5)

$$\int_0^h\left[f(x+t)-\sum_{i=0}^r\frac{t^i}{i!}f_{(i),p}(x)\right] dt = o(h^{r+1}), \text{ as } h\to 0. \qquad (2.13.1)$$

Hence, there is $\delta > 0$ such that

$$\left|\frac{1}{t^{r+1}}\int_0^t\left[f(x+\xi)-\sum_{i=0}^r\frac{\xi^i}{i!}f_{(i),p}(x)\right] d\xi\right| < 1, \text{ for } 0 < |t| < \delta, \qquad (2.13.2)$$

and so

$$\left|\frac{1}{h}\int_0^h\frac{1}{t^{r+1}}\int_0^t\left[f(x+\xi)-\sum_{i=0}^r\frac{\xi^i}{i!}f_{(i),p}(x)\right] d\xi\, dt\right| \le 1, \text{ for } 0 < |h| < \delta.$$

Hence,

$$\left|\int_0^h\frac{1}{t^{r+1}}\int_0^t\left[f(x+\xi)-\sum_{i=0}^r\frac{\xi^i}{i!}f_{(i),p}(x)\right] d\xi\, dt\right| \le |h|, \text{ for } 0 < |h| < \delta. \quad (2.13.3)$$

Let $0 < h < \delta$ and choose ϵ, $0 < \epsilon < h$. Then, on integrating by parts,

$$\int_\epsilon^h \frac{1}{t^r}\left[f(x+t) - \sum_{i=0}^r \frac{t^i}{i!}f_{(i),p}(x)\right]dt = \frac{1}{h^r}\int_\epsilon^h \left[f(x+t) - \sum_{i=0}^r \frac{t^i}{i!}f_{(i),p}(x)\right]dt$$

$$+ r\int_\epsilon^h \frac{1}{t^{r+1}}\int_\epsilon^t \left[f(x+\xi) - \sum_{i=0}^r \frac{\xi^i}{i!}f_{(i),p}(x)\right]d\xi\,dt. \qquad (2.13.4)$$

Now,

$$\int_0^h \frac{1}{t^{r+1}}\int_0^t \left[f(x+\xi) - \sum_{i=0}^r \frac{\xi^i}{i!}f_{(i),p}(x)\right]d\xi\,dt$$

$$= \left(\int_0^\epsilon + \int_\epsilon^h\right)\frac{1}{t^{r+1}}\int_0^t \left[f(x+\xi) - \sum_{i=0}^r \frac{\xi^i}{i!}f_{(i),p}(x)\right]d\xi\,dt$$

$$= \int_\epsilon^h \frac{1}{t^{r+1}}\left(\int_0^\epsilon + \int_\epsilon^t\right)\left[f(x+\xi) - \sum_{i=0}^r \frac{\xi^i}{i!}f_{(i),p}(x)\right]d\xi\,dt$$

$$+ \int_0^\epsilon \frac{1}{t^{r+1}}\int_0^t \left[f(x+\xi) - \sum_{i=0}^r \frac{\xi^i}{i!}f_{(i),p}(x)\right]d\xi\,dt. \qquad (2.13.5)$$

Applying (2.13.2), we have

$$\left|\int_\epsilon^h \frac{1}{t^{r+1}}\int_0^\epsilon \left[f(x+\xi) - \sum_{i=0}^r \frac{\xi^i}{i!}f_{(i),p}(x)\right]d\xi\,dt\right| \le \int_\epsilon^h \frac{\epsilon^{r+1}}{t^{r+1}}\,dt = \frac{\epsilon^{r+1}}{r}\left(\frac{1}{\epsilon^r} - \frac{1}{h^r}\right),$$

which tends to zero as $\epsilon \to 0$ and so,

$$\lim_{\epsilon\to 0}\int_\epsilon^h \frac{1}{t^{r+1}}\int_0^\epsilon \left[f(x+\xi) - \sum_{i=0}^r \frac{\xi^i}{i!}f_{(i),p}(x)\right]d\xi\,dt = 0. \qquad (2.13.6)$$

Applying (2.13.3),

$$\left|\int_0^\epsilon \frac{1}{t^{r+1}}\int_0^t \left[f(x+\xi) - \sum_{i=0}^r \frac{\xi^i}{i!}f_{(i),p}(x)\right]d\xi\,dt\right| \le \epsilon,$$

and so,

$$\lim_{\epsilon\to 0}\int_0^\epsilon \frac{1}{t^{r+1}}\int_0^t \left[f(x+\xi) - \sum_{i=0}^r \frac{\xi^i}{i!}f_{(i),p}(x)\right]d\xi\,dt = 0. \qquad (2.13.7)$$

From (2.13.5), (2.13.6) and (2.13.7),

$$\lim_{\epsilon\to 0}\int_\epsilon^h \frac{1}{t^{r+1}}\int_\epsilon^t \left[f(x+\xi) - \sum_{i=0}^r \frac{\xi^i}{i!}f_{(i),p}(x)\right]d\xi\,dt$$

$$= \int_0^h \frac{1}{t^{r+1}}\int_0^t \left[f(x+\xi) - \sum_{i=0}^r \frac{\xi^i}{i!}f_{(i),p}(x)\right]d\xi\,dt. \qquad (2.13.8)$$

Letting $\epsilon \to 0$ in (2.13.4), we have (using (2.13.8))

$$\int_0^h \frac{1}{t^r}\Big[f(x+t) - \sum_{i=0}^r \frac{t^i}{i!} f_{(i),p}(x)\Big]\,dt = \frac{1}{h^r}\int_0^h \Big[f(x+t) - \sum_{i=0}^r \frac{t^i}{i!} f_{(i),p}(x)\Big]\,dt$$

$$+ r\int_0^h \frac{1}{t^{r+1}}\int_0^t \Big[f(x+\xi) - \sum_{i=0}^r \frac{\xi^i}{i!} f_{(i),p}(x)\Big]\,d\xi\,dt. \tag{2.13.9}$$

If $0 < -h < \delta$, (2.13.9) remains valid. So applying (2.13.1) in the right-hand side of (2.13.9), we have that as $h \to 0$

$$\int_0^h \frac{1}{t^r}\Big[f(x+t) - \sum_{i=0}^r \frac{t^i}{i!} f_{(i),p}(x)\Big]\,dt = \frac{1}{h^r}o(h^{r+1}) + r\int_0^h \frac{1}{t^{r+1}}o(t^{r+1})\,dt = o(h). \tag{2.13.10}$$

This shows that $BD_r f(x)$ exists and $BD_r f(x) = f_{(r),p}(x)$.

To prove that the converse is not true, we exhibit a function f such that $BD_1 f(0)$ exists, but $f_{(1),p}(0)$ does not exist.

Let f, ϕ, ψ and F be functions defined for all x, $0 \le x \le 1$, with $f(0) = \phi(0) = F(0) = 0$ and if $0 < x \le 1$,

$$f(x) = \frac{1}{x}\cos(1/x^3), \quad \phi(x) = x^2 \sin(1/x^3), \quad F(x) = x^3 \sin(1/x^3);$$

further, let $\psi(x) = 3\big(\phi(x) - f(x)\big)$.

Since $F'(x) = \psi(x)$, for all x, ψ is special Denjoy integrable in $[0,1]$. Since ϕ is continuous, it follows that f is special Denjoy integrable. However, f is not Lebesgue integrable. For if it were, then since ϕ is continuous, the function ψ also would be Lebesgue integrable. This implies that F is absolutely continuous and this we now show is not the case.

Let $c_n = \sqrt[3]{\frac{2}{(4n+1)\pi}}$ and $d_n = \sqrt[3]{\frac{1}{2n\pi}}$, $n = 1, 2, \ldots$. Then $\{(c_n, d_n)\}$ is a sequence of disjoint intervals lying in $[0, 1]$. Hence, the series $\sum_1^\infty (d_n - c_n)$ converges, but $\sum_1^\infty |F(d_n) - F(c_n)| = \frac{2}{\pi}\sum_1^\infty \frac{1}{4n+1}$ diverges. This implies that F is not absolutely continuous.

Hence f is not Lebesgue integrable, it follows that for all p, $1 \le p < \infty$, $f \notin \mathcal{L}^p(0,1)$ and this implies $f_{(1),p}(0)$ does not exist.

Now for $x \ne 0$, $\big(x^2 \sin(1/x^3)\big)' = 2x \sin(1/x^3) - (3/x^2)\cos(1/x^3)$ and so for $0 < \epsilon < h$,

$$3\int_\epsilon^h \frac{1}{x^2}\cos\frac{1}{x^3}\,dx = 2\int_\epsilon^h x \sin\frac{1}{x^3}\,dx - h^2 \sin\frac{1}{h^3} + \epsilon^2 \sin\frac{1}{\epsilon^3}. \tag{2.13.11}$$

Since the right-hand side of (2.13.11) tends to a limit as $\epsilon \to 0$, we can let $\epsilon \to 0$ in (2.13.11) to obtain

$$3\int_0^h \frac{1}{x^2}\cos\frac{1}{x^3}\,dx = 2\int_0^h x \sin\frac{1}{x^3}\,dx - h^2 \sin\frac{1}{h^3}, \tag{2.13.12}$$

and so, from the definition of f and from (2.13.12),

$$3 \int_0^h \frac{f(x)}{x} \, dx = 2 \int_0^h x \sin \frac{1}{x^3} \, dx - h^2 \sin \frac{1}{h^3} = o(h), \text{ as } h \to 0.$$

Hence, $BD_1 f(0)$ exists with value 0. □

Theorem 2.13.2 *If f is L^p-bounded of order r at x, then f is Borel bounded of order r at x. The converse is not true.*

□   Since f is L^p-bounded of order r at x, by Corollary 1.12.5 of Chapter I, $f_{(p-1),p}(x)$ exists and

$$\left[\frac{1}{h} \int_0^h \left| f(x+t) - \sum_{i=0}^{r-1} \frac{t^i}{i!} f_{(i),p}(x) \right|^p dt \right]^{1/p} = O(h^r), \text{ as } h \to 0. \qquad (2.13.13)$$

From (2.13.13), we get as in (2.13.1)

$$\int_0^h \left[f(x+t) - \sum_{i=0}^{r-1} \frac{t^i}{i!} f_{(i),p}(x) \right] dt = O(h^{r+1}), \text{ as } h \to 0. \qquad (2.13.14)$$

So, there are $M > 0$ and $\delta > 0$ such that

$$\left| \frac{1}{t^{r+1}} \int_0^t \left[f(x+\xi) - \sum_{i=0}^{r-1} \frac{\xi^i}{i!} f_{(i),p}(x) \right] d\xi \right| < M, \text{ for } 0 < |t| < \delta,$$

and hence,

$$\left| \int_0^h \frac{1}{t^{r+1}} \int_0^t \left[f(x+\xi) - \sum_{i=0}^{r-1} \frac{\xi^i}{i!} f_{(i),p}(x) \right] d\xi \, dt \right| \leq M|h|, \text{ for } 0 < |h| < \delta. \qquad (2.13.15)$$

The relation (2.13.15) is analogous to (2.13.3). Applying arguments similar to those used to deduce (2.13.9) from (2.13.3) and then using (2.13.14) instead of (2.13.1), we get the following result, which is analogous to (2.13.10):

$$\left| \int_0^h \frac{1}{t^r} \left[f(x+t) - \sum_{i=0}^{r-1} \frac{t}{i!} f_{(i),p}(x) \right] dt \right| = O(h), \text{ as } h \to 0. \qquad (2.13.16)$$

Hence, f is Borel bounded of order r at x.
The converse is as in Theorem 2.13.1. □

Corollary 2.13.3 *If f is L^p-bounded of order r at x, then the Borel derivative $BD_{r-1} f(x)$ exists finitely and $\overline{BD}_r f(x)$ and $\underline{BD}_r f(x)$ are finite.*

□   This follows from Theorem 2.13.2 above and Theorem 1.10.1 of Chapter I. □

Theorem 2.13.4 *Let f be special Denjoy integrable in some neighbourhood of x and let F be an indefinite integral of f in that neighbourhood. If $BD_r f(x)$ exists finitely, then $F_{(r+1),p}(x)$ exists with value $BD_r f(x)$.*

☐　　　By Theorem 2.7.4, $F_{(r+1)}(x)$ exists with value $BD_r f(x)$. Since F is an indefinite integral of f, it is continuous and so is in L^p in some neighbourhood of x for all p, $1 \le p < \infty$. So, by Theorem 2.10.1, the result follows.　　　　☐

2.14　Symmetric Borel and Symmetric L^p-Derivatives, $SBD_{(k)}f$ and $f^{(s)}_{(k),p}$

Theorem 2.14.1 *If the symmetric L^p-derivative $f^{(s)}_{(r),p}(x)$ exists, then the symmetric Borel derivative $SBD_r f(x)$ exists with the same value, but the converse is not true.*

☐　　　Let $f^{(s)}_{(r),p}(x)$ exist. Then as $h \to 0$,

$$\left(\frac{1}{h} \int_0^h \left| \frac{1}{2}\left(f(x+t)+(-1)^r f(x-t)\right) - \sum_{i=0}^{m} \frac{t^{r-2i}}{(r-2i)!} f^{(s)}_{(r-2i),p}(x) \right|^p dt \right)^{1/p} = o(h^r),$$

where $m = r/2$ or $(r-1)/2$ according as r is even or odd. Applying Hölder's inequality, we have as in (2.13.1)

$$\int_0^h \left[\frac{1}{2}\left(f(x+t)+(-1)^r f(x-t)\right) - \sum_{i=0}^{m} \frac{t^{r-2i}}{(r-2i)!} f^{(s)}_{(r-2i),p}(x) \right] dt = o(h^{r+1}), \text{ as } h \to 0.$$

So, there is a $\delta > 0$ such that if $0 < |t| < \delta$,

$$\left| \frac{1}{t^{r+1}} \int_0^t \left[\frac{1}{2}\left(f(x+\xi)+(-1)^r f(x-\xi)\right) - \sum_{i=0}^{m} \frac{\xi^{r-2i}}{(r-2i)!} f^{(s)}_{(r-2i),p}(x) \right] d\xi \right| < 1.$$

Hence, for $0 < |h| < \delta$,

$$\left| \int_0^h \frac{1}{t^{r+1}} \int_0^t \left[\frac{1}{2}\left(f(x+\xi)+(-1)^r f(x-\xi)\right) - \sum_{i=0}^{m} \frac{\xi^{r-2i}}{(r-2i)!} f^{(s)}_{(r-2i),p}(x) \right] d\xi \, dt \right|$$
$$\le |h|. \tag{2.14.1}$$

This is similar to that in (2.13.3) and applying arguments similar to those used to deduce (2.13.10) from (2.13.3), we have that as $h \to 0$

$$\int_0^h \frac{1}{t^r} \left[\frac{1}{2}\left(f(x+t)+(-1)^r f(x-t)\right) - \sum_{i=0}^{m} \frac{t^{r-2i}}{(r-2i)!} f^{(s)}_{(r-2i),p}(x) \right] dt = o(h).$$
$$\tag{2.14.2}$$

This shows that $SBD_r f(x)$ exists and $SBD_r f(x) = f^{(s)}_{(r),p}(x)$.

For the converse, consider the function

$$f(x) = \begin{cases} \dfrac{1}{x} \cos(1/x^3), & \text{if } -1 < x \leq 1,\ x \neq 0, \\ 0, & \text{if } x = 0. \end{cases}$$

Then, we have seen in Theorem 2.13.1 that $f \notin \mathcal{L}^p(-1,1)$ and so, $f^{(s)}_{(2),p}(0)$ cannot exist. However, f is odd and, so, $\frac{1}{2}[f(x) + f(-x)] = 0$ for all x, which implies that $SBD_r f(0) = 0$ for all even values of r. □

Theorem 2.14.2 *If f is symmetric L^p-bounded of order r at x, then f is symmetric Borel bounded of order r at x. The converse is not true.*

□ Since f is symmetric L^p-bounded of order r at x, by the relation (2.14.4) and Theorem 1.13.3 of Chapter I, $f^{(s)}_{(r-2),p}(x)$ exists and as $h \to 0$

$$\left(\frac{1}{h} \int_0^h \left| \frac{1}{2}(f(x+t) + (-1)^r f(x-t)) - \sum_{i=1}^m \frac{t^{r-2i}}{(r-2i)!} f^{(s)}_{(r-2i),p}(x) \right|^p dt \right)^{1/p} = O(h^r),$$

where $m = r/2$ or $(r-1)/2$ according as r is even or odd. Then, applying Hölder's inequality, we have as in (2.13.1)

$$\int_0^h \left[\frac{1}{2}(f(x+t) + (-1)^r f(x-t)) - \sum_{i=1}^m \frac{t^{r-2i}}{(r-2i)!} f^{(s)}_{(r-2i),p}(x) \right] dt = O(h^{r+1}), \text{ as } h \to 0.$$

So, there is an $M > 0$ and a $\delta > 0$ such that if $0 < |t| < \delta$,

$$\left| \frac{1}{t^{r+1}} \int_0^t \left[\frac{1}{2}(f(x+\xi) + (-1)^r f(x-\xi)) - \sum_{i=1}^m \frac{\xi^{r-2i}}{(r-2i)!} f^{(s)}_{(r-2i),p}(x) \right] d\xi \right| < M.$$

Hence, for $0 < |h| < \delta$,

$$\left| \int_0^h \frac{1}{t^{r+1}} \int_0^t \left[\frac{1}{2}(f(x+\xi) + (-1)^r f(x-\xi)) - \sum_{i=1}^m \frac{\xi^{r-2i}}{(r-2i)!} f^{(s)}_{(r-2i),p}(x) \right] d\xi\, dt \right|$$
$$\leq M|h|.$$

This is similar to (2.14.1) and applying arguments similar to those used to deduce (2.13.10) from (2.13.3), we have that as $h \to 0$

$$\int_0^h \frac{1}{t^r} \left[\frac{1}{2}(f(x+t) + (-1)^r f(x-t)) - \sum_{i=1}^m \frac{t^{r-2i}}{(r-2i)!} f^{(s)}_{(r-2i),p}(x) \right] dt = O(h).$$

Hence, f is symmetric Borel bounded of order r at x.

The converse is as in Theorem 2.14.1. □

Theorem 2.14.3 *If f is L^p-smooth of order r at x, then f is Borel smooth of order r at x.*

☐   Since f is L^p-smooth of order r at x, $f^{(s)}_{(r-2),p}(x)$ exists and as $h \to 0$

$$\left(\frac{1}{h} \int_0^h \left| \frac{1}{2}\left(f(x+t)+(-1)^r f(x-t)\right) - \sum_{i=1}^m \frac{t^{r-2i}}{(r-2i)!} f^{(s)}_{(r-2i),p}(x) \right|^p dt \right)^{1/p} = o(h^{r-1}),$$

where $m = r/2$ or $(r-1)/2$ according as r is even or odd. Then, proceeding as in Theorem 2.14.1, we get as in (2.14.2)

$$\int_0^h \frac{1}{t^{r-1}} \left[\frac{1}{2}\left(f(x+t)+(-1)^r f(x-t)\right) - \sum_{i=1}^m \frac{t^{r-2i}}{(r-2i)!} f^{(s)}_{(r-2i),p}(x) \right] dt = o(h).$$

(2.14.3)

Now, by Theorem 2.14.1, $SBD_{r-2}f(x)$ exists and $SBD_{r-2i}f(x) = f^{(s)}_{(r-2i),p}(x)$, $i = 1, 2, \ldots, m$, and so, from (2.14.3) and from the definition of $\overline{\varpi}_r(f; x, t)$, (see relation (1.11.6) of Chapter I), we get

$$\int_0^h t\overline{\varpi}_r(f; x, t)\, dt = o(h), \text{ as } h \to 0.$$

Hence, f is Borel smooth of order r at x.   ☐

Theorem 2.14.4 *If $f \in \mathcal{L}^p$, $1 \le p < \infty$, in some neighbourhood of x let F be an indefinite integral of f in that neighbourhood. If $SBD_r f(x)$ exists finitely, then $F^{(s)}_{(r+1),p}(x)$ exists and equals $SBD_r f(x)$. Moreover, if $\overline{SBD}_{r+2}f(x)$ and $\underline{SBD}_{r+2}f(x)$ are finite, then F is L^p-bounded of order $r+3$ at x.*

☐   By Theorem 2.8.2, $F^{(s)}_{(r+1)}(x)$ exists with value $SBD_r f(x)$. So, by Theorem 2.12.1, $F^{(s)}_{(r+1),p}(x)$ exists and equals $F^{(s)}_{(r+1)}(x)$, which gives the first part of the theorem.

For the second part, we have by Theorem 2.8.2 that $\overline{F}^{(s)}_{(r+3)}(x)$ and $\underline{F}^{(s)}_{(r+3)}(x)$ are finite and, so, by Theorem 2.12.5, F is L^p-bounded of order $r+3$ at x.   ☐

2.15   Cesàro and Borel Derivatives, $C_k Df$ and $BD_k f$

Theorem 2.15.1 *Let f be $C_{r-1}P$-integrable in some neighbourhood of x and let Φ be its rth indefinite integral in that neighbourhood. If f is C_r-continuous at x, then*

$$\underline{C_r D}f(x) \le \underline{BD}_{r+1}\Phi(x) \le \overline{BD}_{r+1}\Phi(x) \le \overline{C_r D}f(x)$$ (2.15.1)

and

$$(r+2)\underline{BD}_{r+1}\Phi(x) \; - \; (r+1)\overline{BD}_{r+1}\Phi(x) \le \underline{C_{r+1}D}f(x) \le \overline{C_{r+1}D}f(x)$$
$$\le (r+2)\overline{BD}_{r+1}\Phi(x) - (r+1)\underline{BD}_{r+1}\Phi(x), \qquad (2.15.2)$$

provided that the two extremes of (2.15.2) are defined.

☐   By Theorem 2.4.2, $\Phi_{(r)}(x) = f(x)$ and

$$\underline{C_r D}f(x) = \underline{\Phi}_{(r+1)}(x), \; \overline{C_r D}f(x) = \overline{\Phi}_{(r+1)}(x). \qquad (2.15.3)$$

Also, Φ is a $C_0 P$-integral (see Theorem 2.4.1) and, hence, is continuous. Since $\Phi_{(r)}(x)$ exists finitely, by the last part of Theorem 2.7.1,

$$\underline{\Phi}_{(r+1)}(x) \le \underline{BD}_{r+1}\Phi(x) \le \overline{BD}_{r+1}\Phi(x) \le \overline{\Phi}_{(r+1)}(x). \qquad (2.15.4)$$

The proof of (2.15.1) follows from (2.15.3) and (2.15.4).

To prove (2.15.2), we note that since $\Phi_{(r)}(x)$ exists, $BD_r\Phi(x)$ exists by Theorem 2.7.1. So, by Theorem 2.7.4,

$$(r+2)\underline{BD}_{r+1}\Phi(x) \; - \; (r+1)\overline{BD}_{r+1}\Phi(x) \le \underline{\Psi}_{r+2}(x) \le \overline{\Psi}_{r+2}(x)$$
$$\le (r+2)\overline{BD}_{r+1}\Phi(x) - (r+1)\underline{BD}_{r+1}\Phi(x), \qquad (2.15.5)$$

where Ψ is an indefinite integral of Φ in some neighbourhood of x, and provided the extreme terms in (2.15.5) are defined. Since f is $C_{r-1}P$-integrable, it is $C_r P$-integrable and, so,

$$\frac{h^{r+1}}{(r+1)!}C_{r+1}(f; x, x+h) \; = \; \frac{1}{r!}C_r P - \int_x^{x+h} (x+h-t)^r f(t)\, dt$$

$$= \frac{1}{r!}C_{r-1}P - \int_x^{x+h} (x+h-t)^r f(t)\, dt. \; (2.15.6)$$

Now, using the relation (2.4.1) of Theorem 2.4.1 and integrating by parts successively, we have

$$\frac{1}{r!}\int_x^{x+h} (x+h-t)^r f(t)\, dt \qquad (2.15.7)$$

$$= \frac{1}{r!}(x+h-t)^r F_1(t)\Big|_x^{x+h} + \frac{1}{(r-1)!}\int_x^{x+h} (x+h-t)^{r-1} F_1(t)\, dt$$

$$= -\frac{h^r}{r!}F_1(x) + \frac{1}{(r-1)!}(x+h-t)^{r-1} F_2(t)\Big|_x^{x+h}$$

$$+ \frac{1}{(r-2)!}\int_x^{x+h} (x+h-t)^{r-2} F_2(t)\, dt$$

$$= \cdots\cdots\cdots\cdots\cdots\cdots\cdots\cdots\cdots\cdots\cdots$$

$$= -\sum_{i=0}^{r-2}\frac{h^{r-i}}{(r-i)!}F_{i+1}(x) + (x+h-t)\Phi(t)\Big|_x^{x+h} + \int_x^{x+h}\Phi(t)\, dt$$

$$= -\sum_{i=0}^{r-2}\frac{h^{r-i}}{(r-i)!}F_{i+1}(x) - h\Phi(x) + \Psi(x+h) - \Psi(x),$$

where $F_1, F_2, \ldots$ are as in (2.4.1) of Theorem 2.4.1. Since, by Theorem 2.4.1, $\Phi_{(k)}(x) = F_{r-k}(x)$, $0 < k < r$, from (2.15.6) and (2.15.7)

$$\frac{h^{r+1}}{(r+1)!}C_{r+1}(f; x, x+h) = \Psi(x+h) - \Psi(x) - \sum_{i=0}^{r-1} \frac{h^{r-i}}{(r-i)!}\Phi_{(r-i-1)}(x),$$

or

$$C_{r+1}(f; x, x+h) = \frac{(r+1)!}{h^{r+1}}\left[\Psi(x+h) - \Psi(x) - \sum_{i=1}^{r} \frac{h^i}{i!}\Phi_{(i-1)}(x)\right]. \quad (2.15.8)$$

Since Φ is a C_0P-integral, it is continuous and, hence, by Theorem 2.4.1

$$\lim_{h\to 0} \frac{(r+1)!}{h^{r+1}}\left[\Psi(x+h) - \Psi(x) - \sum_{i=1}^{r} \frac{h^i}{i!}\Phi_{(i-1)}(x)\right]$$

$$= \lim_{h\to 0} \frac{r!}{h^r}\left[\Phi(x+h) - \sum_{i=0}^{r-1} \frac{h^i}{i!}\Phi_i(x)\right] = \lim_{h\to 0} \gamma_r(\Phi; x, h)$$

$$= \lim_{h\to 0} C_r(f; x, x+h) = f(x),$$

since f is C_r-continuous at x. So,

$$\Psi(x+h) - \Psi(x) - \sum_{i=1}^{r} \frac{h^i}{i!}\Phi_{(i-1)}(x) - \frac{h^{r+1}}{(r+1)!}f(x) = o(h^{r+1}), \text{ as } h \to 0,$$

which shows that $\Psi_{(i)}(x) = \Phi_{(i-1)}(x)$, $i = 1, 2, \ldots, r$, and $\Psi_{(r+1)}(x) = f(x)$. Therefore, from (2.15.8), $C_{r+1}(f; x, x+h) = \gamma_{r+1}(\Psi; x, h)$. So,

$$\frac{r+2}{h}\left[C_{r+1}(f; x, x+h) - f(x)\right] = \frac{r+2}{h}\left[\gamma_{r+1}(\Psi; x, h) - f(x)\right] = \gamma_{r+2}(\Psi; x, h),$$

which shows that

$$\underline{C_{r+1}Df}(x) = \underline{\Psi}_{(r+2)}(x), \quad \overline{C_{r+1}D}f(x) = \overline{\Psi}_{(r+2)}(x). \quad (2.15.9)$$

From (2.15.5) and (2.15.9), we get (2.15.2). $\square$

Corollary 2.15.2 *Let f be $C_{r-1}P$-integrable in some neighbourhood of x and let Φ be its rth indefinite integral in that neighbourhood. If f is C_r-continuous at x, then*

(i) *if $C_rDf(x)$ exists, $BD_{r+1}\Phi(x)$ exists with value $C_rDf(x)$;*

(ii) *if $BD_{r+1}\Phi(x)$ exists finitely, then $C_{r+1}Df(x)$ exists and equals $BD_{r+1}\Phi(x)$.*

We now exhibit a function that gives strict inequality in the extreme terms of (2.15.1). Let

$$f(x) = \begin{cases} 2x\sin 1/x^2 - \dfrac{2}{x}\cos 1/x^2, & \text{if } x \neq 0, \\ 0, & \text{if } x = 0; \end{cases} \qquad F(x) = \begin{cases} x^2\sin 1/x^2, & \text{if } x \neq 0, \\ 0, & \text{if } x = 0. \end{cases}$$

Then, $F' = f$ so f is special Denjoy integrable in every neighbourhood of 0 and F is its indefinite integral. Hence,

$$C_1(f;0,h) = \frac{1}{h}\int_o^h f = \frac{F(h)}{h} = h\sin 1/h^2,$$

showing that f is C_1-continuous at 0. Further,

$$\underline{C_1Df}(0) = \liminf_{h\to 0}\frac{2}{h}\big[C_1(f;0,h) - f(0)\big] = -2,$$

and similarly $\overline{C_1D}f(0) = 2$.

Also,

$$(x^3\cos 1/x^2)' = 3x^2\cos 1/x^2 + 2\sin 1/x^2, \ x \neq 0,$$

and so

$$\frac{1}{h}\int_0^h \sin 1/x^2 \, dx = \frac{h^2}{2}\cos 1/h^2 - \frac{3}{2h}\int_0^h x^2\cos 1/x^2 \, dx \to 0 \text{ as } h \to 0.$$

Hence,

$$\frac{1}{h}\int_0^h \frac{F(x)}{x^2} \, dx = \frac{1}{h}\int_0^h \sin 1/x^2 \, dx \to 0 \text{ as } h \to 0;$$

that is, $BD_2F(0) = 0$.

2.16 Symmetric Cesàro and Symmetric Borel Derivatives, SC_kDf and SBD_kf

Theorem 2.16.1 *Let f be $C_{r-1}P$-integrable in some neighbourhood of x and let Φ be its rth indefinite integral in that neighbourhood. Then,*

$$\underline{SC_rDf}(x) \leq \underline{SBD}_{r+1}\Phi(x) \leq \overline{SBD}_{r+1}\Phi(x) \leq \overline{SC_rD}f(x), \qquad (2.16.1)$$

and

$$(r+2)\underline{SBD}_{r+1}\Phi(x) - (r+1)\overline{SBD}_{r+1}\Phi(x) \leq \underline{SC_{r+1}D}f(x)$$
$$\leq \overline{SC_{r+1}D}f(x) \leq (r+2)\overline{SBD}_{r+1}\Phi(x) - (r+1)\underline{SBD}_{r+1}, \qquad (2.16.2)$$

provided the extreme terms in (2.16.2) are defined.

$\Box$ We first observe that the relation $\Phi_{(k)}(x) = F_{r-k}(x)$, $0 < k < r$, still holds even if f is not C_r-continuous at x. For F_1, being a $C_{r-1}P$-integral is C_{r-1}-continuous and $C_{r-1}P$-integrable and, so, applying Theorem 2.4.1 with f and r replaced by F_1 and $r-1$, respectively, we have that $\Phi_{(r-1)}(x) = F_1(x)$ and $\Phi_{(k)}(x) = F_{r-k}(x)$, $0 < k < r$.

The hypotheses of Theorem 2.6.1 imply the existence of $\Phi_{(r-1)}(x)$ in some neighbourhood of x and by Corollary 2.6.2

$$\underline{SC_rD}f(x) = \underline{\Phi}^{(s)}_{(r+1)}(x), \quad \overline{SC_rD}f(x) = \overline{\Phi}^{(s)}_{(r+1)}(x). \tag{2.16.3}$$

Also Φ, being a C_0P-integral, is continuous and so, by the last part of Theorem 2.8.1,

$$\underline{\Phi}^{(s)}_{(r+1)}(x) \leq \underline{SBD}_{r+1}\Phi(x) \leq \overline{SBD}_{r+1}\Phi(x) \leq \overline{\Phi}^{(s)}_{(r+1)}(x). \tag{2.16.4}$$

Hence, we have (2.16.1) from (2.16.3) and (2.16.4).

As for (2.16.2), since $\Phi_{(r-1)}(x)$ exists, $BD_{r-1}\Phi(x)$ exists by Theorem 2.7.1 and so $SBD_{r-1}\Phi(x)$ exists by Theorem 2.9.1. Hence, by the last part of Theorem 2.8.2,

$$(r+2)\underline{SBD}_{r+1}\Phi(x) - (r+1)\overline{SBD}_{r+1}\Phi(x) \leq \underline{\Psi}^{(s)}_{(r+2)}(x) \tag{2.16.5}$$

$$\leq \overline{\Psi}^{(s)}_{(r+2)}(x) \leq (r+2)\overline{SBD}_{r+1}\Phi(x) - (r+1)\underline{SBD}_{r+1},$$

where Ψ is an indefinite integral of Φ in some neighbourhood of x and provided that the extreme terms in (2.16.5) are defined. Since f is $C_{r-1}P$-integrable, f is C_rP-integrable and, so, by the observation made at the beginning of this proof, and by using the argument applied to deduce (2.15.8) from (2.15.6),

$$C_{r+1}(f; x, x+h) = \frac{(r+1)!}{h^{r+1}}\Big[\Psi(x+h) - \Psi(x) - \sum_{i=1}^{r}\frac{h^i}{i!}\Phi_{(i-1)}(x)\Big]. \tag{2.16.6}$$

Since, by the observation, $\Phi_{(r-1)}(x)$ exists,

$$\Phi(x+h) - \Phi(x) - \sum_{i=1}^{r-1}\frac{h^i}{i!}\Phi_{(i)}(x) = o(h^{r-1}), \text{ as } h \to 0,$$

and so, Φ being continuous, we have

$$\lim_{h\to 0}\frac{r!}{h^r}\Big[\Psi(x+h) - \Psi(x) - \sum_{i=1}^{r}\frac{h^i}{i!}\Phi_{(i-1)}(x)\Big]$$

$$= \lim_{h\to 0}\frac{(r-1)!}{h^{r-1}}\Big[\Phi(x+h) - \sum_{i=0}^{r-1}\frac{h^i}{i!}\Phi_{(i)}(x)\Big] = 0.$$

This gives

$$\Psi(x+h) - \Psi(x) - \sum_{i=1}^{r}\frac{h^i}{i!}\Phi_{(i-1)}(x) = o(h^r), \text{ as } h \to 0.$$

Therefore, $\Psi_{(i)}(x) = \Phi_{(i-1)}(x)$, $1 \le i \le r$. So, from (2.16.6),

$$C_{r+1}(f; x, x+h) = \frac{(r+1)!}{h^{r+1}} \left[\Psi(x+h) - \Psi(x) - \sum_{i=1}^{r} \frac{h^i}{i!} \Psi_{(i)}(x)\right] = \gamma_{r+1}(\Psi; x, h).$$

So, by Theorem 2.5.1 (ii)

$$\frac{r+2}{2h} \left[C_{r+1}(f; x, x+h) - C_{r+1}(f; x, x-h)\right] =$$

$$\frac{r+2}{2h} \left[\gamma_{r+1}(\Psi; x, h) - \gamma_{r+1}(\Psi; x, -h)\right] = \varpi(\Psi; x, h),$$

which shows that

$$\underline{SC_{r+1}D}f(x) = \underline{\Psi}_{(r+2)}^{(s)}(x), \quad \overline{SC_{r+1}D}f(x) = \overline{\Psi}_{(r+2)}^{(s)}(x). \tag{2.16.7}$$

The relation (2.16.5) and (2.16.7) imply (2.16.2). $\qquad\square$

Corollary 2.16.2 *Let f be $C_{r-1}P$-integrable in some neighbourhood of x and let Φ be its rth indefinite integral in that neighbourhood. Then,*

(i) *if $SC_rDf(x)$ exists, $SBD_{r+1}\Phi(x)$ exists with value $SC_rDf(x)$;*

(ii) *if $SBD_{r+1}\Phi(x)$ exists finitely, then $SC_{r+1}Df(x)$ exists and $SC_{r+1}Df(x) = SBD_{r+1}\Phi(x)$.*

Theorem 2.16.3 *Let f be $C_{r-1}P$-integrable in some neighbourhood of x and let Φ be its rth indefinite integral in that neighbourhood. If f is SC_r-continuous at x then Φ is Borel smooth of order $r+1$ at x, but the converse is not true.*

$\square$ The proof follows from Corollary 2.6.2 and Theorem 2.8.3. $\qquad\square$

2.17 Abel and Symmetric de la Vallée Poussin Derivatives, $AD_k f$ and $f_{(k)}^{(s)}$

The Abel derivative has a different form and is rather complicated; the arguments used in this section are greatly influenced by Zygmund [196] and Verblunsky [177].

Theorem 2.17.1 *Let f be 2π-periodic and Lebesgue integrable. Then for all x,*

$$\underline{f}_{(1)}^{(s)}(x) \le \underline{AD_1}f(x) \le \overline{AD_1}f(x) \le \overline{f}_{(1)}^{(s)}(x).$$

☐   We prove the right-hand inequality with $x = x_0$; the proof of the left-hand inequality is similar.

There is no loss in generality in assuming $\overline{f}^{(s)}_{(1)}(x_0) < \infty$. Let

$$\frac{1}{2}a_0 + \sum_{n=1}^{\infty}(a_n \cos nx + b_n \sin nx) \tag{2.17.1}$$

be the Fourier series of f and write

$$f(r, x) = \frac{1}{2}a_0 + \sum_{n=1}^{\infty}(a_n \cos nx + b_n \sin nx)r^n, \quad 0 < r < 1. \tag{2.17.2}$$

Then, by (1.14.3) of Chapter I,

$$f(r, x) = \frac{1}{\pi}\int_{-\pi}^{\pi} f(x + t)P(r, t)\, dt = \frac{1}{\pi}\int_{-\pi}^{\pi} f(t)P(r, t - x)\, dt, \tag{2.17.3}$$

where

$$P(r, t) = \frac{1}{2}a_0 + \sum_{n=1}^{\infty} r^n \cos nt = \frac{1}{2}\frac{1 - r^2}{\Delta}, \tag{2.17.4}$$

where $\Delta = \Delta(t) = \Delta(r, t) = 1 - 2r\cos t + r^2$.

Hence, writing $P'(r, x) = (\partial/\partial x)P(r, x)$,

$$P'(r, x) = -\sum_{n=1}^{\infty} nr^n \sin nx = \frac{-r(1 - r^2)\sin x}{\Delta^2(x)}. \tag{2.17.5}$$

In this section, we will use the prime notation exclusively for differentiation with respect to the second variable in P. In what follows, we shall write Δ to mean $\Delta(t)$ unless otherwise stated.

From (2.17.3) and (2.17.5),

$$
\begin{aligned}
\frac{1}{r}\frac{\partial}{\partial x}f(r, x)\Big|_{x=x_0} &= -\frac{1}{\pi r}\int_{-\pi}^{\pi} f(t)P'(r, t - x_0)\, dt \\
&= -\frac{1}{\pi r}\int_{-\pi}^{\pi} f(x_0 + t)P'(r, t)\, dt \\
&= \frac{1}{\pi r}\int_{-\pi}^{\pi} f(x_0 - t)P'(r, t)\, dt \tag{2.17.6} \\
&= -\frac{1}{\pi r}\int_{-\pi}^{\pi} \frac{f(x_0 + t) - f(x_0 - t)}{2}P'(r, t)\, dt \\
&= \frac{1}{\pi}\int_{-\pi}^{\pi} g(t)K(r, t)\, dt,
\end{aligned}
$$

where

$$g(t) = \frac{f(x_0 + t) - f(x_0 - t)}{2\sin t} \quad \text{and} \quad K(r, t) = -\frac{P'(r, t)\sin t}{r} = \frac{(1 - r^2)\sin^2 t}{\Delta^2}.$$

Since $g(t) = g(-t)$ and $K(r,t) = K(r,-t)$, we have from (2.17.6) that

$$\frac{1}{r}\frac{\partial}{\partial x}f(r,x)\Big|_{x=x_0} = \frac{2}{\pi}\int_0^\pi g(t)K(r,t)\,dt. \qquad (2.17.7)$$

Since $\lim\sup_{t\to 0} g(t) = \overline{f}^{(s)}_{(1)}(x_0)$, given an $\epsilon > 0$ there is a $\delta > 0$ such that $g(t) < \overline{f}^{(s)}_{(1)}(x_0) + \epsilon$ if $0 < t < \delta$. Hence, noting that $K(r,t) \geq 0$, $0 \leq t \leq \pi$,

$$\frac{2}{\pi}\int_0^\delta g(t)K(r,t)\,dt \leq \frac{2}{\pi}(\overline{f}^{(s)}_{(1)}(x_0) + \epsilon)\int_0^\delta K(r,t)\,dt$$

$$< \frac{2}{\pi}(\overline{f}^{(s)}_{(1)}(x_0) + \epsilon)\int_0^\pi K(r,t)\,dt. \qquad (2.17.8)$$

The relation (2.17.7) holds for all 2π-periodic Lebesgue integrable functions and for all choices of x_0 and, hence, we can put $f(x) = \sin x$ and $x_0 = 0$ when $g(t) = 1$ for all t and, so, from (2.17.7), $1 = \frac{2}{\pi}\int_0^\pi K(r,t)\,dt$. Using this in (2.17.8),

$$\frac{2}{\pi}\int_0^\delta g(t)K(r,t)\,dt < \overline{f}^{(s)}_{(1)}(x_0) + \epsilon. \qquad (2.17.9)$$

Also,

$$\frac{2}{\pi}\left|\int_\delta^\pi g(t)K(r,t)\,dt\right| \leq \frac{2}{\pi}\sup_{\delta\leq t\leq\pi}K(r,t)\int_\delta^\pi |g(t)|\,dt. \qquad (2.17.10)$$

For $\delta \leq t \leq \pi$, $(\Delta(t))^{-1} \leq (1 - 2r\cos\delta + r^2)^{-1}$ and, so,

$$|P'(r,t)| = \left|\frac{-r(1-r^2)\sin t}{\Delta^2(t)}\right| \leq \frac{r(1-r^2)}{(1-2r\cos\delta + r^2)^2}.$$

Hence, if $\delta \leq t \leq \pi$,

$$\lim_{r\to 1}\sup_{\delta\leq t\leq\pi}|K(r,t)| = \lim_{r\to 1}\sup_{\delta\leq t\leq\pi}\left|\frac{P'(r,t)\sin t}{r}\right|$$

$$\leq \lim_{r\to 1}\sup_{\delta\leq t\leq\pi}\frac{(1-r^2)}{(1-2r\cos\delta + r^2)^2} = 0.$$

Hence, from (2.17.10),

$$\lim_{r\to 1}\frac{2}{\pi}\int_\delta^\pi g(t)K(r,t)\,dt = 0. \qquad (2.17.11)$$

From (2.17.7), (2.17.9) and (2.17.11),

$$\lim_{r\to 1}\sup\frac{1}{r}\frac{\partial}{\partial x}f(r,x)\Big|_{x=x_0} \leq \overline{f}^{(s)}_{(1)}(x_0) + \epsilon,$$

and since ϵ is arbitrary, this proves that

$$\limsup_{r \to 1} \frac{1}{r} \frac{\partial}{\partial x} f(r, x)\Big|_{x=x_0} \le \overline{f}_{(1)}^{(s)}(x_0),$$

which implies that $\overline{AD}_1 f(x_0) \le \overline{f}_{(1)}^{(s)}(x_0)$ as had to be proved.   □

Corollary 2.17.2 *Let f be 2π-periodic and Lebesgue integrable. Then if $f_{(1)}^{(s)}(x)$ exists so does $AD_1 f(x)$, with the same value.*

This result has not been extended to higher order Abel derivatives because inequalities analogous to (2.17.1) are not known. However, a weaker result for the second order Abel derivative can be given. The following result is known as Rajchman's Lemma; see [196] [p. 353] and [177] [p. 445].

Theorem 2.17.3 *Let f be 2π-periodic and Lebesgue integrable with a Fourier series Abel summable at x_0 to $f(x_0)$. Then,*

$$\overline{AD}_2 f(x_0) \ge \underline{f}_{(2)}^{(s)}(x_0) \quad \text{and} \quad \underline{AD}_2 f(x_0) \le \overline{f}_{(2)}^{(s)}(x_0). \qquad (2.17.12)$$

□   We prove the first inequality and the second inequality follows by applying the first inequality to the function $-f$.

Let (2.17.1) be the Fourier series of f and the function $f(r, x)$, the Abel means of f, be as in (2.17.2). Then the hypothesis of the theorem says that $f(x_0) = \lim_{r \to 1} f(r, x_0)$. Further, without loss in generality, we may suppose that $x_0 = 0$, $f(x) = f(-x)$ and $f(0) = 0$.

Suppose, if possible, that $\overline{AD}_2 f(0) < \underline{f}_{(2)}^{(s)}(0)$ and let k be such that $\overline{AD}_2 f(0) < k < \underline{f}_{(2)}^{(s)}(0)$. Put $F(x) = f(x) - k(1 - \cos x)$ when $F(0) = 0$ and $F(x) = F(-x)$. The Abel mean of the Fourier series of F, as in (2.17.2), is

$$F(r, x) = \frac{1}{2} a_0 - k + kr \cos x + \sum_{n=1}^{\infty} (a_n \cos nx + b_n \sin nx) r^n, \ 0 < r < 1.$$

$$(2.17.13)$$

Further, since $f(x_0) = \lim_{r \to 1} f(r, x_0)$, we have that $F(0) = \lim_{r \to 1} F(r, 0)$. Also, $\overline{AD}_2 f(0) - k = \overline{AD}_2 F(0)$ and $\underline{f}_{(2)}^{(s)}(0) - k = \underline{F}_{(2)}^{(s)}(0)$ and, so, by our supposition, $AD_2 F(0) < 0 < \underline{F}_{(2)}^{(s)}(0)$. Let

$$\phi(t) = \frac{F(t) + F(-t) - 2F(0)}{\sin^2 t} = \frac{2F(t)}{\sin^2 t},$$

then since $\underline{F}_{(2)}^{(s)}(0) > 0$, $\liminf_{t \to 0} \phi(t) > 0$. So, choosing $h, 0 < h < \liminf_{t \to 0} \phi(t)$, we find an η, $0 < \eta < \pi/3$, such that $\phi(t) > h$ if $0 < t < \eta$. If

$0 < t < \eta$, then $\cos t > \frac{1}{2}$ and, by (2.17.5) $P'(r,t) < 0$, we have

$$-\int_0^\eta \phi(t) \sin t P'(r,t)\,dt \geq -h \int_0^\eta \sin t P'(r,t)\,dt$$

$$= -h\left[\sin t P(r,t)\big|_0^\eta - \int_0^\eta \cos t P(r,t)\,dt\right]$$

$$\geq -h \sin \eta P(r,\eta) + \frac{h}{2}\int_0^\eta P(r,t)\,dt$$

$$= -h \sin \eta P(r,\eta) + \frac{h}{2}\int_0^\pi P(r,t)\,dt - \frac{h}{2}\int_\eta^\pi P(r,t)\,dt$$

$$= -h \sin \eta P(r,\eta) + \frac{h\pi}{4} - \frac{h}{2}\int_\eta^\pi P(r,t)\,dt.$$

So, using the relations (1.14.5) and (1.14.9) of Chapter I,

$$\liminf_{r\to 1}\left(-\int_0^\eta \phi(t)\sin t P'(r,t)\,dt\right) \geq \frac{h\pi}{4}.$$

Further, by (2.17.5),

$$\liminf_{r\to 1}\left(-\int_\eta^\pi \phi(t)\sin t P'(r,t)\,dt\right) = 0,$$

and so

$$\liminf_{r\to 1}\left(-\int_0^\pi \phi(t)\sin t P'(r,t)\,dt\right) \geq \frac{h\pi}{4}. \qquad (2.17.14)$$

Writing $G(r) = F(r,0)$ and applying (2.17.3) and (2.17.4),

$$G(r) = \frac{1}{\pi}\int_{-\pi}^\pi F(t)P(r,t)\,dt = \frac{1}{\pi}\int_0^\pi F(t)\frac{1-r^2}{\Delta}\,dt.$$

So, differentiating with respect to r and using (2.17.5),

$$\frac{d}{dr}\left(\frac{rG(r)}{1-r^2}\right) = \frac{d}{dr}\left(\frac{1}{\pi}\int_0^\pi F(t)\frac{r}{\Delta}\,dt\right) = \frac{1}{\pi}\int_0^\pi F(t)\frac{1-r^2}{\Delta^2}\,dt$$

$$= \frac{1}{2\pi}\int_0^\pi \phi(t)\sin^2 t\frac{1-r^2}{\Delta^2}\,dt$$

$$= -\frac{1}{2\pi r}\int_0^\pi \phi(t)\sin t P'(r,t)\,dt,$$

hence, from (2.17.14),

$$\liminf_{r\to 1}\frac{d}{dr}\left(\frac{rG(r)}{1-r^2}\right) \geq \frac{h}{8}. \qquad (2.17.15)$$

Now,

$$\frac{d}{dr}\left(\frac{G(r)}{\log r}\right) = \frac{d}{dr}\left(\frac{1-r^2}{r\log r}\frac{rG(r)}{1-r^2}\right)$$

$$= \frac{1-r^2}{r\log r}\frac{d}{dr}\left(\frac{rG(r)}{1-r^2}\right) + \frac{d}{dr}\left(\frac{1-r^2}{r\log r}\right)\frac{rG(r)}{1-r^2}. \quad (2.17.16)$$

Since $\dfrac{1-r^2}{r\log r} \to -2$ and $\dfrac{d}{dr}\left(\dfrac{1-r^2}{r\log r}\right) = O(1-r)$ as $r \to 1$ and noting that $\lim_{r\to 1} G(r) = F(0) = 0$, we get from (2.17.15) and (2.17.16) that

$$\limsup_{r\to 1}\frac{d}{dr}\left(\frac{G(r)}{\log r}\right) < 0.$$

So, there is an r_0, $0 < r_0 < 1$, such that $\dfrac{d}{dr}\left(\dfrac{G(r)}{\log r}\right) < 0$, $r_0 < r < 1$. This implies that $\dfrac{G(r)}{\log r}$ is strictly decreasing on that interval; that is,

$$\frac{G(r)}{\log r} > \frac{G(p)}{\log p}, \text{ if } r_0 < r < p < 1. \quad (2.17.17)$$

Now, from (2.17.13),

$$\frac{\partial^2}{\partial x^2}F(r,x) = -krx\cos x - \sum_{n=1}^{\infty}(a_n \cos nx + b_n \sin nx)n^2 r^n$$

$$= -r\frac{\partial}{\partial r}\left(r\frac{\partial}{\partial r}F(r,x)\right). \quad (2.17.18)$$

Since, by definition, $\overline{AD}_2F(0) = \limsup_{r\to 1}\dfrac{\partial^2}{\partial x^2}F(r,x)\Big|_{x=0} < 0$, (2.17.18) implies that $\liminf_{r\to 1} r\dfrac{\partial}{\partial r}\left(r\dfrac{\partial}{\partial r}F(r,x)\right)\Big|_{x=0} > 0$ and so, $\liminf_{r\to 1} r\dfrac{d}{dr}\left(r\dfrac{d}{dr}G(r)\right) > 0$. So, there is an r_1, $0 < r_1 < 1$, such that $\dfrac{d}{dr}\left(r\dfrac{d}{dr}G(r)\right) > 0$, $r_1 < r < 1$; and so $rG'(r)$ is strictly increasing in $(r_1,1)$. Since $\lim_{r\to 1} G(r) = 0$, for each r, $r_1 < r < 1$, there is, by the mean value theorem, a p, $r < p < 1$, such that $\dfrac{G(r)}{\log r} = p\dfrac{d}{dr}G(r)\Big|_{r=p}$. Hence, for each r, $r_1 < r < 1$, there are p and σ such that

$$\frac{G(r)}{\log r} - \frac{G(p)}{\log p} = p\frac{d}{dr}G(r)\Big|_{r=p} - \sigma\frac{d}{dr}G(r)\Big|_{r=\sigma} < 0, \quad r < p < \sigma < 1. \quad (2.17.19)$$

Relations (2.17.19) and (2.17.17) are contradictory and show that the assumption made is false and thus, the theorem is proved.   □

Theorem 2.17.4 *Let f be 2π-periodic and Lebesgue integrable with a Fourier series Abel summable at x_0 to $f(x_0)$. If*

$$I(h) = \frac{3}{2h^3} \int_0^h \Big(\big(f(x_0 + h) + f(x_0 - h) - 2f(x_0) \big) $$
$$ - \big(f(x_0 + u) + f(x_0 - u) - 2f(x_0) \big) \Big) \cos u \, du, \quad (2.17.20)$$

then

$$\liminf_{h \to 0} I(h) \le \underline{AD}_2 f(x_0) \le \overline{AD}_2 f(x_0) \le \limsup_{h \to 0} I(h).$$

$\square$ As before, we may suppose that $x_0 = 0, f(0) = 0$ and $f(x) = f(-x)$. Then (2.17.20) becomes

$$
\begin{aligned}
I(h) &= \frac{3}{2h^3} \int_0^h \big(2f(h) - 2f(u) \big) \cos u \, du \\
&= \frac{3}{h^3} \left(f(h) \sin h - \int_0^h f(u) \cos u \, du \right) \\
&= 3 \frac{\phi(h)}{h^3}, \quad \text{where} \quad \phi(h) = f(h) \sin h - \int_0^h f(u) \cos u \, du. \quad (2.17.21)
\end{aligned}
$$

Thus, we have to prove

$$\liminf_{t \to 0} 3 \frac{\phi(t)}{t^3} \le \underline{AD}_2 f(0) \le \overline{AD}_2 f(0) \le \limsup_{t \to 0} 3 \frac{\phi(t)}{t^3}. \quad (2.17.22)$$

We prove the left-hand inequality, as the proof of the right-hand inequality then follows by applying the left-hand inequality to the function $-f$. Further, we may (without loss in generality) suppose that $\liminf_{t \to 0} \frac{\phi(t)}{t^3} > -\infty$ and choose a k such that $\liminf_{t \to 0} \frac{\phi(t)}{t^3} > k > -\infty$. Then there is a $\delta > 0, 0 < \delta < \pi$, such that $\frac{\phi(t)}{t^3} > k$ if $0 < t < \delta$. So, using notation introduced above,

$$\int_0^\delta \frac{\sin t}{\Delta^3} \phi(t) \, dt = \int_0^\delta \frac{t^3 \sin t}{\Delta^3} \frac{\phi(t)}{t^3} \, dt \ge k \int_0^\delta \frac{t^3 \sin t}{\Delta^3} \, dt.$$

Therefore,

$$
\begin{aligned}
\int_0^\pi \frac{\sin t}{\Delta^3} \phi(t) \, dt &= \left(\int_0^\delta + \int_\delta^\pi \right) \frac{\sin t}{\Delta^3} \phi(t) \, dt \\
&\ge k \int_0^\delta \frac{t^3 \sin t}{\Delta^3} \, dt + \int_\delta^\pi \frac{\sin t}{\Delta^3} \phi(t) \, dt \\
&= k \left(\int_0^\pi - \int_\delta^\pi \right) \frac{t^3 \sin t}{\Delta^3} \, dt + \int_\delta^\pi \frac{\sin t}{\Delta^3} \phi(t) \, dt. \quad (2.17.23)
\end{aligned}
$$

Since,

$$(1-r^2)\left|\int_\delta^\pi \frac{t^3 \sin t}{\Delta^3} \, dt\right| \le (1-r^2)\int_\delta^\pi \frac{t^3 \sin t}{\Delta^3(r,\delta)} \, dt \to 0 \text{ as } r \to 1$$

and

$$(1-r^2)\left|\int_\delta^\pi \frac{\sin t}{\Delta^3} \phi(t) \, dt\right| \le \frac{1-r^2}{\Delta^3(r,\delta)}\int_\delta^\pi |\phi(t)| \, dt \to 0 \text{ as } r \to 1,$$

we have from (2.17.23)

$$\liminf_{r\to 1}(1-r^2)\int_0^\pi \frac{\sin t}{\Delta^3}\phi(t) \, dt \ge \liminf_{r\to 1}(1-r^2)k\int_0^\pi \frac{t^3 \sin t}{\Delta^3} \, dt. \qquad (2.17.24)$$

We now evaluate the right-hand side of (2.17.24).

Integrating by parts and using (2.17.23),

$$\int_0^\pi tP'(r,t) \, dt = tP(r,t)\Big|_{t=0}^{t=\pi} - \int_0^\pi P(r,t) \, dt = \frac{\pi(1-r^2)}{2(1+r)^2} - \frac{\pi}{2} = \frac{-\pi r}{1+r},$$

and so, using (2.17.5)

$$\begin{aligned}
r(1-r^2)\int_0^\pi \frac{t^2}{\Delta^2} \, dt &= \int_0^\pi \left(\frac{-t^2}{\sin t}\right)\frac{-r(1-r^2)\sin t}{\Delta^2} \, dt \\
&= \int_0^\pi \left(t - \frac{t^2}{\sin t}\right)P'(r,t) \, dt - \int_0^\pi tP'(r,t) \, dt \quad (2.17.25) \\
&= \int_0^\pi g(t)P'(r,t) \, dt + \frac{\pi r}{1+r}, \text{ where } g(t) = t - \frac{t^2}{\sin t}.
\end{aligned}$$

Then $g'(t) = 1 - \dfrac{2t\sin t - t^2\cos t}{\sin^2 t} = 1 - \dfrac{t}{\sin t}\left(2 - \dfrac{t}{\sin t}\cos t\right)$. So $g(t) \to 0$ and $g'(t) \to 0$ as $t \to 0$. Let $\epsilon > 0$ be arbitrary and choose $\delta > 0$ such that $|g'(t)| < \epsilon$ if $0 < t < \delta$. So, integrating by parts,

$$\begin{aligned}
\left|\int_0^\delta g(t)P'(r,t) \, dt\right| &= \left|g(t)P(r,t)\Big|_{t=0}^{t=\delta} - \int_0^\delta g'(t)P(r,t) \, dt\right| \\
&\le |g(\delta)P(r,\delta)| + \left|\int_0^\delta g'(t)P(r,t) \, dt\right| \\
&\le |g(\delta)|P(r,\delta) + \epsilon\int_0^\pi P(r,t) \, dt \\
&= |g(\delta)|P(r,\delta) + \epsilon\frac{\pi}{2}.
\end{aligned}$$

Hence,

$$\lim_{r\to 1}\left|\int_0^\delta g(t)P'(r,t) \, dt\right| \le |g(\delta)| \lim_{r\to 1} P(r,\delta) + \epsilon\frac{\pi}{2} = \epsilon\frac{\pi}{2}. \qquad (2.17.26)$$

Also,

$$\lim_{r \to 1} \int_\delta^\pi g(t)P'(r,t)\,dt = \lim_{r \to 1} \int_\delta^\pi \left(t - \frac{t^2}{\sin t}\right)\frac{-r(1-r^2)\sin t}{\Delta^2}\,dt$$

$$= \lim_{r \to 1} r(1-r^2)\int_\delta^\pi \frac{t^2 - t\sin t}{\Delta^2}\,dt = 0. \quad (2.17.27)$$

From (2.17.26) and (2.17.27),

$$\lim_{r \to 1}\int_0^\pi g(t)P'(r,t)\,dt \le \epsilon\frac{\pi}{2},$$

and, since ϵ is arbitrary,

$$\lim_{r \to 1}\int_0^\pi g(t)P'(r,t)\,dt = 0. \quad (2.17.28)$$

Now integrating by parts,

$$\int_0^\pi \frac{8r^2(1-r^2)t^3\sin t}{\Delta^3}\,dt = 4r(1-r^2)\int_0^\pi t^3\frac{2r\sin t}{\Delta^3}\,dt$$

$$= 4r(1-r^2)\left[t^3\int_0^t \frac{2r\sin\xi}{\Delta^3(r,\xi)}\,d\xi\Big|_{t=0}^{t=\pi} - 3\int_0^\pi t^2\int_0^t \frac{2r\sin\xi}{\Delta^3(r,\xi)}\,d\xi\,dt\right]$$

$$= 4r(1-r^2)\left[t^3\frac{-1}{2\Delta^2(r,t)}\Big|_{t=0}^{t=\pi} + \frac{3}{2}\int_0^\pi \frac{t^2}{\Delta^2(r,t)}\,dt\right]$$

$$= \frac{-2r(1-r^2)\pi^3}{(1+r)^4} + 6r(1-r^2)\int_0^\pi \frac{t^2}{\Delta^2(r,t)}\,dt.$$

Letting $r \to 1$ and using (2.17.25) and (2.17.28),

$$\lim_{r \to 1}\int_0^\pi \frac{8r^2(1-r^2)t^3\sin t}{\Delta^3}\,dt = \lim_{r \to 1} 6\frac{r\pi}{1+r} = 3\pi. \quad (2.17.29)$$

Using (2.17.29), we get from (2.17.24),

$$\liminf_{r \to 1}(1-r^2)\int_0^\pi \frac{\sin t}{\Delta^3}\,dt \ge \frac{3\pi k}{8}. \quad (2.17.30)$$

It can be verified by differentiating (2.17.2) and (2.17.4) that

$$\frac{\partial^2}{\partial x^2}f(r,x) = \frac{1}{\pi}\int_{-\pi}^\pi f(t)P''(r,t)\,dt. \quad (2.17.31)$$

Now, from (2.17.4),

$$P''(r,x) = \frac{-r(1-r^2)\cos x}{\Delta(x)^2} + \frac{2r(1-r^2)\sin x\,2r\sin x}{\Delta^3(x)}$$

$$= r(1-r^2)\left[\frac{4r\sin^2 x}{\Delta(x)^3} - \frac{\cos x}{\Delta^2(x)}\right],$$

and, so, from (2.17.31),

$$\frac{\partial^2}{\partial x^2} f(r,x)\Big|_{x=x_0} \tag{2.17.32}$$

$$= \frac{1}{\pi} \int_0^\pi f(t) \left(2r(1-r^2) \left(\frac{4r\sin^2 t}{\Delta^3} - \frac{\cos t}{\Delta^2} \right) \right) dt$$

$$= \frac{1}{\pi} \int_0^\pi \frac{8r^2(1-r^2)\sin t}{\Delta^3} f(t)\sin t\, dt - \frac{1}{\pi} \int_0^\pi \frac{2r(1-r^2)}{\Delta^2} f(t)\cos t\, dt.$$

Integrating by parts,

$$\int_0^\pi \frac{1}{\Delta^2} f(t)\cos t\, dt \tag{2.17.33}$$

$$= \frac{1}{\Delta^2} \int_0^t f(\xi)\cos\xi\, d\xi \Big|_{t=0}^{t=\pi} + \int_0^\pi \frac{2.2r\sin t}{\Delta^3} \left(\int_0^t f(\xi)\cos\xi\, d\xi \right) dt$$

$$= \frac{1}{(1+r)^2} \int_0^\pi f(\xi)\cos\xi\, d\xi + \int_0^\pi \frac{4r\sin t}{\Delta^3} \left(\int_0^t f(\xi)\cos\xi\, d\xi \right) dt.$$

From (2.17.32) and (2.17.33),

$$\frac{\partial^2}{\partial x^2} f(r,x)\Big|_{x=x_0} =$$

$$\frac{1}{\pi} \int_0^\pi \frac{8r^2(1-r^2)\sin t}{\Delta^3} f(t)\sin t\, dt - \frac{2r(1-r^2)}{\pi} \left[\frac{1}{(1+r)^2} \int_0^\pi f(t)\cos t\, dt \right.$$

$$\left. + \int_0^\pi \frac{4r\sin t}{\Delta^3} \left(\int_0^t f(\xi)\cos\xi\, d\xi \right) dt \right]$$

$$= -\frac{2r(1-r^2)}{\pi(1+r)^2} \int_0^\pi f(t)\cos t\, dt + \frac{8r^2(1-r^2)}{\pi} \int_0^\pi \frac{\sin t}{\Delta^3} \left(f(t)\sin t \right.$$

$$\left. - \int_0^t f(\xi)\cos\xi\, d\xi \right) dt$$

$$= -\frac{2r(1-r^2)}{\pi(1+r)^2} \int_0^\pi f(t)\cos t\, dt + \frac{8r^2(1-r^2)}{\pi} \int_0^\pi \frac{\sin t}{\Delta^3} \phi(t)\, dt,$$

where $\phi(t)$ is defined in (2.17.21). So, using (2.17.30),

$$\liminf_{r\to 1} \frac{\partial^2}{\partial x^2} f(r,x)\Big|_{x=x_0} \geq 3k.$$

Since k was arbitrary, we can conclude

$$\underline{AD}_2 f(0) \geq 3 \liminf_{t\to 0} \frac{\phi(t)}{t^3},$$

proving the left-hand inequality in (2.17.22).   □

Theorem 2.17.5 *Let f be 2π-periodic and Lebesgue integrable, with*
$\lim_{t\to 0} \dfrac{f(x_0 + t) + f(x_0 - t)}{2} = f(x_0)$. *If*

$$J(h) = \frac{3}{2h^3} \int_0^h \Big(\big(f(x_0 + h) + f(x_0 - h) - 2f(x_0) \big)$$
$$- \big(f(x_0 + u) + f(x_0 - u) - 2f(x_0) \big) \Big) \, du, \qquad (2.17.34)$$

then

$$\liminf_{h\to 0} J(h) \le \underline{AD}_2 f(x_0) \le \overline{AD}_2 f(x_0) \le \limsup_{h\to 0} J(h).$$

□ We show that $\lim_{h\to 0} \big| J(h) - I(h) \big| = 0$, where $I(h)$ is the integral defined in (2.17.20) and the result follows from Theorem 2.17.4 because by Lemma 2.15.2 of Chapter I, the Fourier series of f is Abel summable at x_0 to $f(x_0)$.

As before, we may suppose that $x_0 = 0$, $f(0) = 0$ and $f(x) = f(-x)$ when

$$\big| J(h) - I(h) \big| = \frac{3}{2h^3} \left| \int_0^h (f(h) - f(u))(1 - \cos u) \, du \right|.$$

Let $\epsilon > 0$ be arbitrary, then since $f(t) \to f(0) = 0$ and $(1 - \cos t)/t^2 \to 1/2$ as $t \to 0$, there is a $\delta > 0$ such that $\big| f(t) \big| < \epsilon$ and $|1 - \cos t| < (\epsilon + 1/2)t^2$ if $0 < t < \delta$. Then, from the above,

$$\big| J(h) - I(h) \big| \le \frac{3}{2h^3} \int_0^h \big| f(h) - f(u) \big| (1 - \cos u) \, du$$
$$\le \frac{3}{2h^3} \left[\int_0^h \big| f(h) \big| (1 - \cos u) \, du + \int_0^h \big| f(u) \big| (1 - \cos u) \, du \right]$$
$$\le \frac{3}{2h^3} \left[\epsilon \left(\frac{1}{2} + \epsilon \right) \frac{h^3}{3} + \epsilon \left(\frac{1}{2} + \epsilon \right) \frac{h^3}{3} \right] = \frac{\epsilon}{2}(1 + 2\epsilon).$$

Hence, $\lim_{h\to 0} \big| J(h) - I(h) \big| = 0$. □

Theorem 2.17.6 *Let f be 2π-periodic and Lebesgue integrable and let*
$\lim_{t\to 0} \dfrac{f(x_0 + t) + f(x_0 - t)}{2} = f(x_0)$. *Then,*

$$\frac{3}{2} f_{-(2)}^{(s)}(x_0) - \frac{1}{2} \overline{f}_{(2)}^{(s)}(x_0) \le \underline{AD}_2 f(x_0) \le \overline{AD}_2 f(x_0) \le \frac{3}{2} \overline{f}_{(2)}^{(s)}(x_0) - \frac{1}{2} f_{-(2)}^{(s)}(x_0),$$
$$(2.17.35)$$

provided the extreme terms of these inequalities are defined.

□ We will only consider the the left-hand inequality as once this is proved the right-hand inequality is obtained by applying the left-hand inequality to the function $-f$.

Note that, in particular, the extreme terms are not defined if $f_{(2)}^{(s)}(x_0) = \pm\infty$. So, we may assume that $\overline{f}_{(2)}^{(s)}(x_0) < \infty$ for if it is ∞, then $f_{-(2)}^{(s)}(x_0) < \infty$

and the left-hand side is $-\infty$ and the result is obvious. If $\overline{f}^{(s)}_{(2)}(x_0) < \infty$ and $\underline{f}^{(s)}_{(2)}(x_0) = -\infty$, the result is again obvious, thus, we may assume that both derivates are finite. In addition, we may assume without loss in generality that $x_0 = 0$, $f(0) = 0$ and $f(x) = f(-x)$. Choose then k_1, k_2 such that $k_2 < \underline{f}^{(s)}_{(2)}(0) \le \overline{f}^{(s)}_{(2)}(0) < k_1$. Then there exists $\delta > 0$ such that

$$k_2 < \frac{f(x_0 + h) + f(x_0 - h) - 2f(x_0)}{h^2} = \frac{2f(h)}{h^2} < k_1, \text{ for } 0 < h < \delta.$$

So, if $0 < h < \delta$, we have

$$\frac{3}{h^3} \int_0^h 2f(u)\,du \le \frac{3}{h^3} k_1 \int_0^h u^2\,du = k_1$$

and

$$\frac{3}{h^3} \int_0^h 2f(h)\,du = \frac{3}{h^3} 2f(h)h > 3k_2.$$

So,

$$\liminf_{h \to 0} \frac{3}{2h^3} \int_0^h \left(2f(h) - 2f(u)\right) du \ge \liminf_{h \to 0} \frac{3}{h^3} \int_0^h f(h)\,du$$

$$- \limsup_{h \to 0} \frac{3}{h^3} \int_0^h f(u)\,du$$

$$\ge \frac{3}{2} k_2 - \frac{1}{2} k_1. \tag{2.17.36}$$

Thus, by Theorem 2.17.5, $\underline{AD}_2 f(0) \ge \frac{3}{2} k_2 - \frac{1}{2} k_1$ and, since k_1 and k_2 are arbitrary, this gives the left-hand inequality in (2.17.35). □

Corollary 2.17.7 *Let f be 2π-periodic and Lebesgue integrable. If $f^{(s)}_{(2)}(x_0)$ exists finitely, then $AD_2 f(x_0)$ exists with value $f^{(s)}_{(2)}(x_0)$.*

Remark. Another proof of this corollary is in Lemma 2.9.2 of [196] [356].

Analogues of Theorem 2.17.3 and Corollary 2.17.7 are not known for higher order Abel and d.l.V.P. derivates. However, a slightly less general result is the following:

Theorem 2.17.8 *Let f be 2π-periodic and Lebesgue integrable. If the upper and lower d.l.V.P. derivates $\overline{f}^{(s)}_{(r)}(x_0)$ and $\underline{f}^{(s)}_{(r)}(x_0)$ are finite, then there is a $K > 0$ such that*

$$-K\lambda \le \underline{AD}_r f(x_0) \le \overline{AD}_r f(x_0) \le K\lambda,$$

where $\lambda = \max\{|\overline{f}^{(s)}_{(r)}(x_0)|, |\underline{f}^{(s)}_{(r)}(x_0)|\}$. Moreover, if $f^{(s)}_{(r)}(x_0)$ exists finitely, then $AD_r f(x_0)$ exists and equals $f^{(s)}_{(r)}(x_0)$.

□   This result is proved in [121, Theorem 2.1 and Corollary] when r is even, and [121, Theorem 2.2] when r is odd.   □

Theorem 2.17.9 *Let f be 2π-periodic and Lebesgue integrable with a Fourier series Abel summable at x_0 to $f(x_0)$. Then,*

$$\underline{S}_2 f(x_0) \leq \pi \overline{AS}_2 f(x_0), \quad \overline{S}_2 f(x_0) \geq \pi \underline{AS}_2 f(x_0).$$

□   We prove the first inequality by showing that if for any k, $\underline{S}_2 f(x_0) > k$, then $\pi \overline{AS}_2 f(x_0) \geq k$; the proof of the second is similar. We may suppose that $x_0 = 0$, $f(x_0) = 0$ and $f(x) = f(-x)$.

Define the 2π-periodic function $g(t)$ by letting $g(t) = \frac{1}{2}k|t|$, $-\pi < t \leq \pi$, where k is as above. Then, as in (2.17.31),

$$
\begin{aligned}
\frac{\partial^2}{\partial x^2} g(r,x)\Big|_{x=0} &= \frac{1}{\pi}\int_{-\pi}^{\pi} g(t) P''(r,t)\,dt = \frac{2}{\pi}\int_0^{\pi} k\frac{t}{2} P''(r,t)\,dt \\
&= \frac{k}{\pi}\left[tP'(r,t)\Big|_{t=0}^{t=\pi} - \int_0^{\pi} P'(r,t)\,dt \right] \\
&= -\frac{k}{\pi} P(r,t)\Big|_{t=0}^{t=\pi}, \quad \text{by (2.17.5)}, \\
&= \frac{k}{\pi}\frac{1}{2}\left[\frac{1-r^2}{(1-r)^2} - \frac{1-r^2}{(1+r)^2} \right], \quad \text{by (2.17.4)}, \\
&= \frac{k}{\pi(1-r)}\frac{2r}{1+r}.
\end{aligned}
$$

Hence, $\lim_{r\to 1-} \pi(1-r)\dfrac{\partial^2}{\partial x^2} g(r,x)\Big|_{x=0} = k$. This shows that the theorem holds for the function g and $x_0 = 0$. As a result, we may subtract g from f and it is sufficient to prove that if $\underline{S}_2 f(0) > 0$, then $\pi\overline{AS}_2 f(0) \geq 0$, or just that $\overline{AS}_2 f(0) \geq 0$.

Suppose then that $\underline{S}_2 f(0) > 0$ when $\lim_{h\to 0+} f(h)/h > 0$. Hence, here is a $\delta > 0$ such that $f(t) > 0$ for $0 < t < \delta$. We may assume that $f > 0$ everywhere. For, since

$$
\begin{aligned}
\frac{\partial^2}{\partial x^2} f(r,x)\Big|_{x=0} &= \frac{1}{\pi}\int_0^{\pi} f(t) P''(r,t)\,dt \\
&= \frac{1}{\pi}\left(\int_0^{\delta} + \int_{\delta}^{\pi} \right) f(t) P''(r,t)\,dt
\end{aligned}
$$

and

$$\lim_{r\to 1-}(1-r)\int_{\delta}^{\pi} f(t) P''(r,t)\,dt = \lim_{r\to 1-}(1-r)\int_{\delta}^{\pi} f(t)\frac{1-r^2}{2}\frac{\partial^2}{\partial t^2}\left(\frac{1}{\Delta}\right)dt = 0,$$

we have

$$\pi\overline{AS}_2 f(0) = \limsup_{r\to 1-} \pi(1-r)\frac{\partial^2}{\partial x^2} f(r,x)\Big|_{x=0} = \limsup_{r\to 1-}(1-r)\int_0^{\delta} f(t) P''(r,t)\,dt.$$

Therefore, the value of $\overline{AS}_2 f(0)$ does not depend on the values of f outside $(0, \delta)$. So,

$$f(r, 0) = \frac{1}{\pi} \int_{-\pi}^{\pi} f(t) P(r, t) \, dt = \frac{1}{\pi} \int_0^\pi f(t) \frac{1 - r^2}{\Delta} \, dt > 0, \text{ for } 0 < r < 1.$$

$$(2.17.37)$$

Since $f(r, x)$ is given by (2.17.2), $f(r, 0) = a_0/2 + \sum_{n=1}^\infty a_n r^n$ and, thus,

$$-r \frac{d}{dr} \left(r \frac{d}{dr} f(r, 0) \right) = -\sum_{n=1}^\infty n^2 a_n r^n = \frac{\partial^2}{\partial r^2} f(r, x) \Big|_{x=0}. \qquad (2.17.38)$$

Now, suppose if possible that $\overline{AS}_2 f(0) < 0$. Then

$$\limsup_{r \to -1} (1 - r) \left(\partial^2 f(r, x) / \partial x^2 \right) \Big|_{x=0} < 0$$

and so there is a $c > 0$ and a σ, $0 < \sigma < 1$, such that

$$\frac{\partial^2}{\partial x^2} f(r, x) \Big|_{x=0} < \frac{-c}{1 - r}, \text{ for } 1 - \sigma < r < 1.$$

Hence, from (2.17.39),

$$\frac{d}{dr} \left(r \frac{d}{dr} f(r, 0) \right) > \frac{c}{1 - r}, \text{ for } 1 - \sigma < r < 1. \qquad (2.17.39)$$

Let $1 - \sigma < r_1 < r_2 < 1$. Then, from (2.17.40),

$$r \frac{d}{dr} f(r, 0) \Big|_{r=r_2} - r \frac{d}{dr} f(r, 0) \Big|_{r=r_1} \geq c \int_{r_1}^{r_2} \frac{1}{1 - r} \, dr$$

$$= c \log \frac{1 - r_1}{1 - r_2} \to \infty \text{ as } r_2 \to 1-.$$

Hence, $\lim_{r_2 \to 1-} r \frac{d}{dr} f(r, 0) \Big|_{r=r_2} = \infty$ and, so, $f(r, 0)$ is strictly increasing in some left neighbourhood of 1. Since the Fourier series of f is Abel summable at 0 to $f(0) = 0$, $f(r, 0) \to 0$ as $r \to -1$. It follows that $f(r, 0)$ is negative in some left neighbourhood of 1. However, this contradicts (2.17.38) and so our assumption that $\overline{AS}_2 f(0) < 0$ is false and, so, $\overline{AS}_2 f(0) \geq 0$, as had to be proved. $\square$

Just as Theorem 2.17.9 is the analogue of Theorem 2.17.3 for 2-smoothness, a theorem is proved in [122, Theorem 2.1 and Theorem 2.2] and is an analogue of Theorem 2.17.8 for r-smoothness.

We now give two examples. The first will show that the converse of Corollary 2.17.7 does not hold; the second will show that the use of the Lebesgue integral in this theory is necessary.

Example 1. Let f be 2π-periodic and

$$f(x) = \begin{cases} \dfrac{\pi - x}{2}, & \text{if } 0 < x \leq 2\pi \text{ and irrational}, \\ 1, & \text{if } 0 < x \leq 2\pi \text{ and rational}, \\ 0, & \text{if } x = 0. \end{cases}$$

Then, $\sum_{n=1}^{\infty} \sin nx / n$ is the Fourier series of f and, thus, $f(r, x) = \sum_{n=1}^{\infty} (r^n \sin nx)/n$. Hence,

$$\frac{\partial^2}{\partial x^2} f(r, x) = -\sum_{n=1}^{\infty} nr^n \sin nx = P'(r, x) = \frac{-r(1 - r^2) \sin x}{(1 - 2r \cos x + r^2)^2}.$$

Therefore, $AD^2 f(x) = 0$ for all x; but $f_{(2)}^{(s)}(x)$ does not exist for any x.

This same example shows that $AD^2 f(x) = 0$ for all x does not imply that f is a polynomial of degree at most, unless other conditions are satisfied 2.

Example 2. If

$$F(x) = \begin{cases} x^2 \cos 1/x^2, & \text{if } x \neq 0, \\ 0, & \text{if } x = 0, \end{cases}$$

then F is differentiable. Define f as a 2π-periodic function with

$$f(x) = \begin{cases} F'(x), & \text{if } 0 \leq x \leq \pi, \\ f(-x), & \text{if } -\pi \leq x \leq 0. \end{cases}$$

Since F' is not Lebesgue integrable on $[0, \pi]$, f is not Lebesgue integrable on $[-\pi, \pi]$ and, thus, the existence of $AD_k f(x)$ cannot arise for any k and any x.

Note that although F' is special Denjoy integrable and, hence, f is also special Denjoy integrable, the Fourier Denjoy coefficients of f need not tend to zero and, therefore, the definition of $AD_k f$ fails.

However, $f^{(k)}(x)$ exists for all k and for all $x, x \neq 0$.

Remark. Theorem 2.17.3 is due to Rajchman and Zygmund, see [177] [p. 445] and [196] [p. 353]. Theorems 2.17.4 and 2.17.5 are due to Verblunsky; see [177] [pp. 446–447]. Theorem 2.17.9 is proved in [177] [p. 443] and also in [196] [p. 357]. The proofs given in these references are related to Fourier series. We have modified and simplified these proofs to serve our purpose.

2.18 Laplace, Peano and Generalized Peano Derivatives, $LD_k f$, $f_{(k)}$ and $f_{[k]}$

Lemma 2.18.1 *Let f be special Denjoy integrable in $I = [a, b]$ and C_1-continuous at $x_0 \in I$. If $\ell \in \mathbb{N}$, write $f^{(-\ell)}$ for the ℓ-th indefinite integral of f*

on I. Then, for each $k \in \mathbb{N}$

$$(f^{(-k)})^{(r)}(x_0) = f^{(-k+r)}(x_0), \text{ for } r = 1, 2, \ldots, k. \tag{2.18.1}$$

☐   Since $f^{(-1)}$ is an indefinite Denjoy integral of f on I and since f is C_1-continuous at x_0,

$$\frac{f^{(-1)}(x_0 + t) - f^{(-1)}(x_0)}{t} = \frac{1}{t} \int_{x_0}^{x_0+t} f \to f(x_0) \text{ as } t \to 0. \tag{2.18.2}$$

Since $f^{(-1)}$ is continuous in I,

$$(f^{(-k)})^{(r)}(x_0) = f^{(-k+r)}(x_0), \text{ for } r = 1, 2, \ldots, k-1, \tag{2.18.3}$$

and, from (2.18.2), $(f^{(-k)})^{(k)}(x_0) = (f^{(-1)})^{(1)}(x_0) = f(x_0)$. This together with (2.18.3) gives (2.18.1).   ☐

Theorem 2.18.2 *Let f be special Denjoy integrable and C_1-continuous in $I = [a, b]$. If $n \in \mathbb{N}$, then for all $k \in \mathbb{N}$, $LD_{n+k}f^{(-k)}(x)$ exists if and only if $LD_nf(x)$ exists, in which case they are equal. More generally, for all k*

$$\overline{LD}^{+}_{n+k}f^{(-k)}(x) = \overline{LD}^{+}_{n}f(x), \quad \underline{LD}^{+}_{n+k}f^{(-k)}(x) = \underline{LD}^{+}_{n}f(x), \tag{2.18.4}$$

with similar relations for the left-hand Laplace derivates.

☐   Let n and k be fixed. By Lemma 2.18.1, the kth derivative of $f^{(-k)}$ at x exists. Let $\alpha_j = (f^{(-k)})^{(j)}(x)$, $j = 0, 1, 2, \ldots, k$. If $n > 1$, choose arbitrarily $\alpha_{k+1}, \ldots, \alpha_{n+k-1}$; if $n = 1$, there is nothing to choose. Integrating by parts,

$$s^{n+k+1} \int_0^\delta e^{-st} \left(f^{(-k)}(x+t) - \sum_{i=0}^{n+k-1} \frac{t^i}{i!} \alpha_i \right) dt$$

$$= s^{n+k+1} \left[\frac{e^{-st}}{-s} \left(f^{(-k)}(x+t) - \sum_{i=0}^{n+k-1} \frac{t^i}{i!} \alpha_i \right) \right] \Bigg|_{t=0}^{t=\delta}$$

$$+ s^{n+k} \int_0^\delta e^{-st} \left(f^{(-k+1)}(x+t) - \sum_{i=1}^{n+k-1} \frac{t^{i-1}}{(i-1)!} \alpha_i \right) dt. \tag{2.18.5}$$

The first term on the right-hand side of (2.18.5) is $o(1)$ as $s \to \infty$ and so, as $s \to \infty$,

$$s^{n+k+1} \int_0^\delta e^{-st} \left(f^{(-k)}(x+t) - \sum_{i=0}^{n+k-1} \frac{t^i}{i!} \alpha_i \right) dt$$

$$= o(1) + s^{n+k} \int_0^\delta e^{-st} \left(f^{(-k+1)}(x+t) - \sum_{i=1}^{n+k-1} \frac{t^{i-1}}{(i-1)!} \alpha_i \right) dt. \tag{2.18.6}$$

In a similar manner we can also prove

$$s^{n+k} \int_0^\delta e^{-st} \left(f^{(-k+1)}(x+t) - \sum_{i=1}^{n+k-1} \frac{t^{i-1}}{(i-1)!} \alpha_i \right) dt \qquad (2.18.7)$$

$$= o(1) + s^{n+k-1} \int_0^\delta e^{-st} \left(f^{(-k+2)}(x+t) - \sum_{i=2}^{n+k-1} \frac{t^{i-2}}{(i-2)!} \alpha_i \right) dt.$$

From (2.18.6) and (2.18.7), we have

$$s^{n+k+1} \int_0^\delta e^{-st} \left(f^{(-k)}(x+t) - \sum_{i=0}^{n+k-1} \frac{t^i}{i!} \alpha_i \right) dt \qquad (2.18.8)$$

$$= o(1) + s^{n+k-1} \int_0^\delta e^{-st} \left(f^{(-k+2)}(x+t) - \sum_{i=2}^{n+k-1} \frac{t^{i-2}}{(i-2)!} \alpha_i \right) dt.$$

As we get (2.18.8) from (2.18.6) using (2.18.7), we get, continuing this process,

$$s^{n+k+1} \int_0^\delta e^{-st} \left(f^{(-k)}(x+t) - \sum_{i=0}^{n+k-1} \frac{t^i}{i!} \alpha_i \right) dt \qquad (2.18.9)$$

$$= o(1) + s^{n+1} \int_0^\delta e^{-st} \left(f(x+t) - \sum_{i=k}^{n+k-1} \frac{t^{i-k}}{(i-k)!} \alpha_i \right) dt$$

$$= o(1) + s^{n+1} \int_0^\delta e^{-st} \left(f(x+t) - \sum_{i=0}^{n-1} \frac{t^i}{i!} \alpha_{i+k} \right) dt.$$

Now, suppose that $LD_{n+k} f^{(-k)}(x)$ exists finitely. Then taking $\alpha_i = LD_i^+ f^{(-k)}(x)$, $i = k+1, k+2, \ldots, n+k-1$, the left-hand side of (2.18.9) tends to $LD_{n+k}^+ f^{(-k)}(x)$ as $s \to \infty$ and, so, the right-hand side of (2.18.9) tends to $LD_n^+ f(x)$ with equal value. If $LD_n^+ f(x)$ exists finitely, then taking $\alpha_{i+k} = LD_i^+ f(x)$, $i = 1, 2, \ldots, n-1$, the right-hand side of (2.18.9) tends to $LD_n^+ f(x)$ as $s \to \infty$ and, so, the left-hand side of (2.18.9) tends to $LD_{n+k}^+ f^{(-k)}(x)$.

To prove the last part: It follows from the definitions of $\overline{LD}_{n+k}^+ f^{(-k)}(x)$ and $\overline{LD}_n^+ f(x)$ that $LD_{n+k-1}^+ f^{(-k)}(x)$ and $LD_{n-1}^+ f(x)$ exist finitely and, so, from the first part, are equal. So, writing $\alpha_i = LD_i^+ f^{(-k)}(x)$, $i = 0, 1, \ldots, n+k-1$, in (2.18.9) and letting $s \to \infty$, we get (2.18.4). $\qquad\square$

Theorem 2.18.3 *Let f be special Denjoy integrable in $I = [a,b]$ and $x \in I$. If $f_{(n-1)}(x)$ exists finitely, then*

$$\underline{f}_{(n)}^+(x) \le \underline{LD}_n^+ f(x) \le \overline{LD}_n^+ f(x) \le \overline{f}_{(n)}^+(x). \qquad (2.18.10)$$

□   We prove the right-hand inequality only as the proof of the left-hand inequality is similar.

If $\overline{f}^{+}_{(n)}(x) = \infty$, there is nothing to prove; so, we may suppose that $\overline{f}^{+}_{(n)}(x) = M < \infty$. Choose an $M_1 < \infty$ such that $M < M_1$. Then there is a $\delta > 0$ such that

$$f(x+t) - \sum_{i=0}^{n-1} \frac{t^i}{i!} f_{(i)}(x) < M_1 \frac{t^n}{n!}, \text{ for } 0 < t < \delta.$$

Therefore, using Lemma 1.15.1 of Chapter I,

$$s^{n+1} \int_0^{\delta} e^{-st} \left(f(x+t) - \sum_{i=0}^{n-1} \frac{t^i}{i!} f_{(i)}(x) \right) dt \le s^{n+1} \frac{M_1}{n!} \int_0^{\delta} e^{-st} t^n \, dt$$

$$= \frac{M_1}{n!} n! s^{n+1-n-1} + o(1) = M_1 + o(1), \text{ as } s \to \infty.$$

So, letting $s \to \infty$, $\overline{LD}^{+}_n f(x) \le M_1$ and, thus, $\overline{LD}^{+}_n f(x) \le M$ as M_1 was arbitrary.   □

Theorem 2.18.4 *Let f be special Denjoy integrable and C_1-continuous in $I = [a, b]$. If $f^{+}_{[n-1]}(x)$ exists finitely, $x \in I$, then*

$$\underline{f}^{+}_{[n]}(x) \le \underline{LD}^{+}_n f(x) \le \overline{LD}^{+}_n f(x) \le \overline{f}^{+}_{[n]}(x).$$

□   Since $f^{+}_{[n-1]}(x)$ exists finitely, there is a $k \in \mathbb{N}$ such that the kth indefinite integral of f in I, $f^{(-k)}$, satisfies $\overline{(f^{(-k)})}^{+}_{(n+k)}(x) = \overline{f}^{+}_{[n]}(x)$ and, by Theorem 2.18.3, $\overline{LD}^{+}_{n+k}(f^{(-k)})(x) \le \overline{(f^{(-k)})}^{+}_{(n+k)}(x)$. So, applying Theorem 2.18.2, we get that $\overline{LD}_n f(x) \le \overline{f}^{+}_{[n]}(x)$.

The other inequality follows similarly.   □

From Theorems 17.3 and 17.4 we get:

Theorem 2.18.5 *Let f be special Denjoy integrable in $I = [a, b]$. Then,*

(i) *if $f_{(n)}(x)$ exists, possibly infinite, $x \in I$, then $LD_n f(x)$ exists with the same value;*

(ii) *if f is C_1-continuous in I and if $f^{+}_{[n]}(x)$ exists, possibly infinite, $x \in I$, then $LD_n f(x)$ exists with the same value.*

The following example shows that the converse is not true.

Example. Let

$$f(x) = \begin{cases} x^n, & \text{if either } x \text{ is irrational or } x = 0, \\ 1, & \text{if } x \text{ is rational, } x \ne 0. \end{cases}$$

Then, $f'(0)$ does not exist. However, for any r, $1 \le r \le n$,

$$s^{r+1} \int_0^\delta e^{-st} f(t)\, dt = s^{r+1} \int_0^\delta e^{-st} t^n\, dt$$

$$= n!\, s^{r+1-n-1} + o(1) = n!\, s^{r-n} + o(1), \text{ as } s \to \infty.$$

So,

$$\lim_{s \to \infty} s^{r+1} \int_0^\delta e^{-st} f(t)\, dt = \begin{cases} n!, & \text{if } r = n, \\ 0, & \text{if } r < n. \end{cases}$$

Hence, $LD_n f(0)$ exists with value $n!$.

2.19 Laplace and Borel Derivatives, $LD_k f$ and $BD_k f$

Theorem 2.19.1 *Let f be special Denjoy integrable in some neighbourhood of x. If $BD_r^+ f(x)$ exists finitely, so does $LD_r^+ f(x)$ with the same value. Moreover,*

$$(r+2)\underline{BD}_{r+1}^+ f(x) - (r+1)\overline{BD}_{r+1}^+ f(x) \le \underline{LD}_{r+1}^+ f(x)$$
$$\le \overline{LD}_{r+1}^+ f(x) \le (r+2)\overline{BD}_{r+1}^+ f(x) - (r+1)\underline{BD}_{r+1}^+ f(x), \quad (2.19.1)$$

provided the extreme terms have meaning. Similar relations hold for the other derivates.

☐   Let $BD_r^+ f(x)$ exist finitely and let $\epsilon > 0$, $\sigma > 0$ be arbitrary. Then, there is a $\delta > 0$ such that

$$\left| \int_0^h \frac{f(x+t) - \sum_{i=0}^r \frac{t^i}{i!} BD_i^+ f(x)}{t^r}\, dt \right| < \epsilon h, \text{ for } 0 < h < \delta. \quad (2.19.2)$$

So, if $0 < \delta_1 < \min\{\delta, \sigma\}$, integrating by parts and using (2.19.2),

$$\left| \int_0^{\delta_1} e^{-st} \left(f(x+t) - \sum_{i=0}^r \frac{t^i}{i!} BD_i^+ f(x) \right) dt \right|$$

$$= \left| \int_0^{\delta_1} e^{-st} t^r \frac{f(x+t) - \sum_{i=0}^r \frac{t^i}{i!} BD_i^+ f(x)}{t^r}\, dt \right|$$

$$\le \left| e^{-st} t^r \int_0^t \frac{f(x+\xi) - \sum_{i=0}^r \frac{\xi^i}{i!} BD_i^+ f(x)}{\xi^r}\, d\xi \right|_{t=0}^{t=\delta_1}$$

$$+ \left| \int_0^{\delta_1} (se^{-st} t^r - re^{-st} t^{r-1}) \int_0^t \frac{f(x+\xi) - \sum_{i=0}^r \frac{\xi^i}{i!} BD_i^+ f(x)}{\xi^r}\, d\xi\, dt \right|$$

$$< e^{-s\delta_1}\delta_1^r \epsilon \delta_1 + s\int_0^{\delta_1} e^{-st} t^r \epsilon t\, dt + r\int_0^{\delta_1} e^{-st} t^{r-1} \epsilon t\, dt$$

$$= e^{-s\delta_1}\delta_1^{r+1}\epsilon + s\epsilon\int_0^{\delta_1} e^{-st} t^{r+1}\, dt + r\epsilon\int_0^{\delta_1} e^{-st} t^r\, dt.$$

So, by Lemma 1.15.1 of Chapter I,

$$\left| s^{r+1}\int_0^{\delta_1} e^{-st}\left(f(x+t) - \sum_{i=0}^r \frac{t^i}{i!} BD_i^+ f(x)\right) dt\right|$$

$$\leq s^{r+1} e^{-s\delta_1}\delta_1^{r+1}\epsilon + \epsilon(r+1)! s^{r+2-r-1-1} + r\epsilon r! s^{r+1-r-1} + o(1),$$

$$\text{as } s\to\infty.$$

Also, it can be shown using integration by parts that

$$\left| s^{r+1}\int_{\delta_1}^\sigma e^{-st}\left(f(x+t) - \sum_{i=0}^r \frac{t^i}{i!} BD_i^+ f(x)\right) dt\right| = o(1) \text{ as } s\to\infty.$$

Thus, letting $s\to\infty$, since ϵ is arbitrary,

$$\lim_{s\to\infty} s^{r+1}\int_0^\sigma e^{-st}\left(f(x+t) - \sum_{i=0}^r \frac{t^i}{i!} BD_i^+ f(x)\right) dt = 0.$$

Hence, $LD_r^+ f(x)$ exists finitely and $LD_r^+ f(x) = BD_r^+ f(x)$, which proves the first part of the theorem.

For the second part, we prove the right-hand inequality; the proof of the left-hand inequality is similar. If $\overline{BD}_{r+1}^+ f(x) = \infty$, then since the right-hand side of (2.19.1) is well defined, $\underline{BD}_{r+1}^+ f(x) \neq \infty$, giving the right-hand side of (2.19.1) the value ∞ and the inequality is obvious. So, we may suppose that $\overline{BD}_{r+1}^+ f(x) < \infty$. Similarly, if $\underline{BD}_{r+1}^+ f(x) = -\infty$, we must have $\overline{BD}_{r+1}^+ f(x) > -\infty$ and again the inequality is obvious. So, we need only consider the case $-\infty < \underline{BD}_{r+1}^+ f(x) \leq \overline{BD}_{r+1}^+ f(x) < \infty$. Then, choose m, M such that

$$-\infty < m < \underline{BD}_{r+1}^+ f(x) \leq \overline{BD}_{r+1}^+ f(x) < M < \infty. \tag{2.19.3}$$

Then, there is $\delta > 0$ such that

$$m < \frac{(r+1)!}{h}\int_0^h \frac{f(x+t) - \sum_{i=0}^r \frac{t^i}{i!} BD_i^+ f(x)}{t^{r+1}}\, dt < M, \text{ for } 0 < h < \delta.$$

$$\tag{2.19.4}$$

Choose $\sigma > 0$ arbitrary and let $0 < \delta_1 < \min\{\delta, \sigma\}$. By the first part, $LD_i^+ f(x)$ exists and equals $BD_i^+ f(x)$, $i = 0, 1, \ldots, r$, and, so, we get, integrating by parts and applying (2.19.4),

$$\int_0^{\delta_1} e^{-st}\left(f(x+t) - \sum_{i=0}^{r} \frac{t^i}{i!} BD_i^+ f(x)\right) dt \tag{2.19.5}$$

$$= \int_0^{\delta_1} e^{-st} t^{r+1} \frac{f(x+t) - \sum_{i=0}^{r} \frac{t^i}{i!} BD_i^+ f(x)}{t^{r+1}} \, dt$$

$$= e^{-st} t^{r+1} \int_0^t \frac{f(x+\xi) - \sum_{i=0}^{r} \frac{\xi^i}{i!} BD_i^+ f(x)}{\xi^{r+1}} \, d\xi \Big|_{t=0}^{t=\delta_1}$$

$$+ \int_0^{\delta_1} \left((se^{-st}t^{r+1} - (r+1)e^{-st}t^r)\int_0^t \frac{f(x+\xi) - \sum_{i=0}^{r} \frac{\xi^i}{i!} BD_i^+ f(x)}{\xi^{r+1}} \, d\xi\right) dt$$

$$= e^{-s\delta_1} \delta_1^{r+1} \int_0^{\delta_1} \frac{f(x+\xi) - \sum_{i=0}^{r} \frac{\xi^i}{i!} BD_i^+ f(x)}{\xi^{r+1}} \, d\xi$$

$$+ s \int_0^{\delta_1} e^{-st} t^{r+1} \int_0^t \frac{f(x+\xi) - \sum_{i=0}^{r} \frac{\xi^i}{i!} BD_i^+ f(x)}{\xi^{r+1}} \, d\xi \, dt$$

$$- (r+1) \int_0^{\delta_1} e^{-st} t^r \int_0^t \frac{f(x+\xi) - \sum_{i=0}^{r} \frac{\xi^i}{i!} BD_i^+ f(x)}{\xi^{r+1}} \, d\xi \, dt$$

$$< \frac{e^{-s\delta_1} \delta_1^{r+2} M}{(r+1)!} + \frac{sM}{(r+1)!} \int_0^{\delta_1} e^{-st} t^{r+2} \, dt - \frac{m}{r!} \int_0^{\delta_1} e^{-st} t^{r+1} \, dt.$$

From (2.19.5) and Lemma 1.15.1 of Chapter I,

$$s^{r+2} \int_0^{\delta_1} e^{-st}\left(f(x+t) - \sum_{i=0}^{r} \frac{t^i}{i!} BD_i^+ f(x)\right) dt$$

$$< s^{r+2} \frac{e^{-s\delta_1} \delta_1^{r+2} M}{(r+1)!} + (r+2)M - (r+1)m + o(1), \text{ as } s \to \infty.$$

Also,

$$s^{r+2} \int_{\delta_1}^{\sigma} e^{-st}\left(f(x+t) - \sum_{i=0}^{r} \frac{t^i}{i!} BD_i^+ f(x)\right) dt = o(1) \text{ as } s \to \infty.$$

So, letting $s \to \infty$, $\overline{LD}_{r+1}^+ f(x) \le (r+2)M - (r+1)m$. Since M and m are arbitrary, this gives the right-hand inequality in (2.19.1). □

Theorem 2.19.2 *If f is special Denjoy integrable in some neighbourhood of x and if f is Borel bounded of order r at x, then f is Laplace bounded of order r at x.*

□ By Theorem 1.10.1 of Chapter I, the Borel derivative of order $r-1$ at x exists finitely and the Borel derivates of order r at x are finite. So, by Theorem

2.19.1, the Laplace derivative of order $r-1$ at x exists finitely and the Laplace derivates of order r at x are finite. So, by Theorem 1.15.3 of Chapter I, f is Laplace bounded of order r at x. □

2.20 Symmetric Laplace and Symmetric de la Vallée Poussin Derivatives, $SLD_k f$ and $f^{(s)}_{(k)}$

Theorem 2.20.1 *Let f be special Denjoy integrable in some neighbourhood of x. If the d.l.V.P. derivative of f at x of order r, $f^{(s)}_{(r)}(x)$, exists finitely, then the symmetric Laplace derivative of f at x of order r, $SLD_r f(x)$, exists and is equal to $f^{(s)}_{(r)}(x)$. Moreover,*

$$\underline{f}^{(s)}_{(r+2)}(x) \leq \underline{SLD}_{r+2}f(x) \leq \overline{SLD}_{r+2}f(x) \leq \overline{f}^{(s)}_{(r+2)}(x). \qquad (2.20.1)$$

□   Let $f^{(s)}_{(r)}(x)$ exist finitely. Then the derivative $f^{(s)}_{(k)}(x)$ exists, $k = r, r - 2, \ldots, 0$ or 1, according as r is even or odd. Write

$$\tau(t) = \sum_{i=0}^{m} \frac{t^{r-2i}}{(r-2i)!} f^{(s)}_{(r-2i)}(x), \qquad (2.20.2)$$

where $m = r/2$ or $(r-1)/2$ according as r is even or odd. Let $\epsilon > 0$, $\sigma > 0$ be arbitrary. Then there is a $\delta > 0$ such that

$$\left| \frac{f(x+t) + (-1)^r f(x-t)}{2} - \tau(t) \right| < \epsilon \frac{t^r}{r!}, \text{ for } 0 \leq t < \delta.$$

Let $0 < \delta_1 < \min\{\delta, \sigma\}$. Then, by Lemma 1.15.1 of Chapter I,

$$\left| s^{r+1} \int_0^{\delta_1} e^{-st} \left(\frac{f(x+t) + (-1)^r f(x-t)}{2} - \tau(t) \right) dt \right| \leq \frac{\epsilon}{r!} s^{r+1} \int_0^{\delta_1} e^{-st} t^r \, dt$$
$$= \epsilon + o(1), \text{ as } s \to \infty.$$

Also,

$$s^{r+1} \int_{\delta_1}^{\sigma} e^{-st} \left(\frac{f(x+t) + (-1)^r f(x-t)}{2} - \tau(t) \right) dt = o(1), \text{ as } s \to \infty.$$

So, letting $s \to \infty$,

$$\lim_{s \to \infty} \left| s^{r+1} \int_0^{\sigma} e^{-st} \left(\frac{f(x+t) + (-1)^r f(x-t)}{2} - \tau(t) \right) dt \right| \leq \epsilon.$$

Hence, since ϵ was arbitrary,

$$s^{r+1} \int_0^\sigma e^{-st} \left(\frac{f(x+t) + (-1)^r f(x-t)}{2} - \tau(t) \right) dt = o(1), \text{ as } s \to \infty.$$

Further, $\tau(t) = \left(\tau(t) + (-1)^r \tau(-t) \right)/2$ and, so, $SLD_r f(x)$ exists with value $f_{(r)}^{(s)}(x)$.

To prove the second part of the theorem, it is sufficient to consider the right-hand inequality and to assume that $\overline{f}_{(r+2)}^{(s)}(x) < \infty$.

So, choose an M such that $\overline{f}_{(r+2)}^{(s)}(x) < M < \infty$ and $\sigma > 0$. Then, there is a $\delta > 0$ such that

$$\frac{f(x+t) + (-1)^r f(x-t)}{2} - \tau(t) \leq M \frac{t^{r+2}}{(r+2)!}, \text{ for } 0 \leq t < \delta.$$

Then, if $0 < \delta_1 < \min\{\delta, \sigma\}$,

$$s^{r+3} \int_0^{\delta_1} e^{-st} \left(\frac{f(x+t) + (-1)^r f(x-t)}{2} - \tau(t) \right) dt$$

$$\leq s^{r+3} \int_0^{\delta_1} e^{-st} M \frac{t^{r+2}}{(r+2)!} dt$$

$$= M + o(1), \text{ as } s \to \infty.$$

And, so, as before,

$$s^{r+3} \int_0^\sigma e^{-st} \left(\frac{f(x+t) + (-1)^r f(x-t)}{2} - \tau(t) \right) dt = M + o(1), \text{ as } s \to \infty.$$

Therefore, letting $s \to \infty$, $\overline{SLD}_{r+2} f(x) \leq M$ and since M was arbitrary, this proves that $\overline{SLD}_{r+2} f(x) \leq \overline{f}_{(r+2)}^{(s)}(x)$. $\qquad \square$

Theorem 2.20.2 *Let f be special Denjoy integrable in some neighbourhood of x. If the d.l.V.P. derivative of f order r, $f_{(r)}^{(s)}(x)$, exists finitely, then*

$$\frac{1}{r+2} \underline{S}_{r+2} f(x) \leq \underline{SL}_{r+2} f(x) \leq \overline{SL}_{r+2} f(x) \leq \frac{1}{r+2} \overline{S}_{r+2} f(x). \qquad (2.20.3)$$

(The definitions of the indices of smoothness in (2.20.3) can be found in Sections 6.2 and 16.3 of Chapter I.)

$\square$ We consider the right-hand inequality in (2.20.3) and assume without loss in generality that $\overline{S}_{r+2} f(x) < M < \infty$. Then, there is a $\delta > 0$ such that

$$\frac{f(x+t) + (-1)^r f(x-t)}{2} - \tau(t) < M \frac{t^{r+1}}{(r+2)!}, \text{ for } 0 \leq t < \delta,$$

where $\tau(t)$ is as defined in (2.20.2). Let $\sigma > 0$ be arbitrary. Then for $0 < \delta_1 < \min\{\delta, \sigma\}$,

$$s^{r+2} \int_0^{\delta_1} e^{-st} \left(\frac{f(x+t) + (-1)^r f(x-t)}{2} - \tau(t) \right) dt$$

$$\leq s^{r+2} \int_0^{\delta_1} e^{-st} M \frac{t^{r+1}}{(r+2)!} dt$$

$$= \frac{M}{r+2} + o(1), \quad \text{as } s \to \infty$$

and, hence,

$$s^{r+2} \int_0^{\sigma} e^{-st} \left(\frac{f(x+t) + (-1)^r f(x-t)}{2} - \tau(t) \right) dt \leq \frac{M}{r+2} + o(1), \quad \text{as } s \to \infty.$$

So, letting $s \to \infty$, and since by Theorem 2.20.2, $SLD_k f(x)$ exists with value $f_{(k)}^{(s)}(x)$, $k = r, r-2, \ldots$, we have $\overline{SL}_{r+2} f(x) \leq M/(r+2)$. Since, then, M is arbitrary, the proof is complete. $\qquad\square$

Corollary 2.20.3 *Under the hypotheses of Theorem 2.20.2, if f is d.l.V P. smooth of order $r+2$ at x, then f is Laplace smooth of order $r+2$ at x.*

Theorem 2.20.4 *If f is d.l.V.P. bounded of order $r+2$ at x, then f is Laplace bounded of order $r+2$ at x.*

$\square$ Since f is d.l.V.P. bounded of order $r+2$ at x, then $f_{(r)}^{(s)}(x)$ exists finitely and, by Theorem 1.6.3 of Chapter I, $\overline{f}_{(r+2)}^{(s)}(x)$ and $\underline{f}_{(r+2)}^{(s)}(x)$ are finite. Hence, there is an M and a $\delta > 0$ such that

$$\left| \frac{f(x+t) + (-1)^r f(x-t)}{2} - \tau(t) \right| < M \frac{t^{r+2}}{(r+2)!}, \quad \text{for } 0 \leq t < \delta,$$

where $\tau(t)$ is as in (2.20.2). Let $\sigma > 0$ be fixed. Then for $0 < \delta_1 < \min\{\delta, \sigma\}$,

$$\left| s^{r+3} \int_0^{\delta_1} e^{-st} \left(\frac{f(x+t) + (-1)^r f(x-t)}{2} - \tau(t) \right) dt \right|$$

$$\leq s^{r+3} \int_0^{\delta_1} e^{-st} M \frac{t^{r+2}}{(r+2)!} dt$$

$$= M + o(1), \quad \text{as } s \to \infty.$$

So,

$$s^{r+3} \int_0^{\sigma} e^{-st} \left(\frac{f(x+t) + (-1)^r f(x-t)}{2} - \tau(t) \right) dt = M + o(1) \text{ as } s \to \infty,$$

which completes the proof. $\qquad\square$

2.21 Laplace and Symmetric Laplace Derivatives, $LD_k f$ and $SLD_k f$

Theorem 2.21.1 *Let f be special Denjoy integrable in some neighbourhood of x. If the Laplace derivative of f of order r at x, $LD_r f(x)$, exists finitely, then the symmetric Laplace derivative of order r at x, $SLD_r f(x)$, exists with the same value. Moreover,*

$$\frac{1}{2}\left(\underline{LD^+_{r+1}f(x)} + \underline{LD^-_{r+1}f(x)}\right) \le \underline{SLD}_{r+1}f(x)$$

$$\le \overline{SLD}_{r+1}f(x) \le \frac{1}{2}\left(\overline{LD}^+_{r+1}f(x) + \overline{LD}^-_{r+1}f(x)\right), \quad (2.21.1)$$

provided the extreme terms are well defined.

☐   Let $LD_r f(x)$ exist finitely and define

$$Q(t) = \sum_{i=0}^{r} \frac{t^i}{i!} LD_i f(x). \quad (2.21.2)$$

Then,

$$s^{r+1} \int_0^\delta e^{-st}\big(f(x+t) - Q(t)\big)\, dt = o(1), \text{ as } s \to \infty \quad (2.21.3)$$

and

$$s^{r+1} \int_0^\delta e^{-st}\big(f(x-t) - Q(-t)\big)\, dt = o(1), \text{ as } s \to \infty. \quad (2.21.4)$$

From (2.21.3) and (2.21.4), as $s \to \infty$

$$s^{r+1} \int_0^\delta e^{-st}\left(\frac{f(x+t) + (-1)^r f(x-t)}{2} - \frac{Q(t) + (-1)^r Q(-t)}{2}\right) dt = o(1).$$

Hence, $SLD_r f(x)$ exists with value $LD_r f(x)$.

To prove (2.21.1), we may suppose that $\overline{LD}^+_{r+1}f(x) < \infty$ and $\overline{LD}^-_{r+1}f(x) < \infty$. Choose M_1 and M_2 such that $\overline{LD}^+_{r+1}f(x) < M_1 < \infty$ and $\overline{LD}^-_{r+1}f(x) < M_2 < \infty$. Then, there exists an $N > 0$ such that

$$s^{r+2} \int_0^\delta e^{-st}\big(f(x+t) - Q(t)\big)\, dt < M_1, \text{ for } s > N \quad (2.21.5)$$

and

$$(-1)^{r+1}s^{r+2} \int_0^\delta e^{-st}\big(f(x-t) - Q(-t)\big)\, dt < M_2, \text{ for } s > N. \quad (2.21.6)$$

From (2.21.5) and(2.21.6), if $s > N$,

$$s^{r+2} \int_0^\delta e^{-st} \left(\frac{f(x+t) + (-1)^{r+1} f(x-t)}{2} - \frac{Q(t) + (-1)^{r+1} Q(-t)}{2} \right) dt$$
$$< \frac{M_1 + M_2}{2}.$$

Letting $s \to \infty$ $\overline{SLD}_{r+1} f(x) \le (M_1 + M_2)/2$ and then letting $M_1 \to \overline{LD}_{r+1}^+ f(x)$ and $M_2 \to \overline{LD}_{r+1}^- f(x)$, the right-hand side of (2.21.1) follows.
　　The left-hand side is proved similarly.　　　　　　　　　　□

Theorem 2.21.2 *If $LD_{r+1} f(x)$ exists finitely, then f is Laplace smooth at x of order $r + 2$.*

□　　Let $LD_{r+1} f(x)$ exist finitely and let $\epsilon > 0$ be arbitrary. Then there exists an $N > 0$ such that

$$\left| s^{r+2} \int_0^\delta e^{-st} \left(f(x+t) - P(t) \right) dt \right| < \epsilon, \quad \text{for } s > N \tag{2.21.7}$$

and

$$\left| s^{r+2} \int_0^\delta e^{-st} \left(f(x-t) - P(-t) \right) dt \right| < \epsilon, \quad \text{for } s > N, \tag{2.21.8}$$

where

$$P(t) = \sum_{i=0}^{r+1} \frac{t^i}{i!} LD_i f(x).$$

From (2.21.7) and (2.21.8), if $s > N$,

$$\left| s^{r+2} \int_0^\delta e^{-st} \left(\frac{f(x+t) + (-1)^r f(x-t)}{2} - \frac{P(t) + (-1)^r P(-t)}{2} \right) dt \right| < \epsilon.$$

Hence, f is Laplace smooth of order $r + 2$.　　　　　　　　□

2.22　Peano and Unsymmetric Riemann Derivatives, $f_{(k)}$ and $RD_k f$

Theorem 2.22.1 *If $f_{(k)}(x)$ exists finitely, then $RD_k f(x)$ exists and is equal to $f_{(k)}(x)$. Further, if $\overline{f}_{(k+1)}(x)$ and $\underline{f}_{(k+1)}(x)$ are finite, then $\overline{RD}_{k+1} f(x)$ and $\underline{RD}_{k+1} f(x)$ are also finite, but not conversely.*

☐   The following relation will be used in the proof.

$$\sum_{i=0}^{k}(-1)^{k-i}\binom{k}{i}i^s = \begin{cases} 0, & \text{if } s = 0, 1, \ldots, k-1, \\ k!, & \text{if } s = k. \end{cases} \tag{2.22.1}$$

Since $f_{(k)}(x)$ exists finitely,

$$f(x+t) = \sum_{i=0}^{k}\frac{t^i}{i!}f_{(i)}(x) + o(t^k), \text{ as } t \to 0. \tag{2.22.2}$$

From (2.22.1) and (2.22.2) (see (2.3.16) and (2.3.21) of Chapter I),

$$\begin{aligned}
\Delta_k(f, x, t) &= \sum_{i=0}^{k}(-1)^{k-i}\binom{k}{i}f(x+it) \\
&= \sum_{i=0}^{k}(-1)^{k-i}\binom{k}{i}\left(\sum_{j=0}^{k}\frac{(it)^j}{j!}f_{(j)}(x) + o(t^k)\right) \\
&= \sum_{j=0}^{k}\frac{t^j}{j!}f_{(j)}(x)\left(\sum_{i=0}^{k}(-1)^{k-i}\binom{k}{i}i^j\right) + o(t^k) \\
&= t^k f_{(k)}(x) + o(t^k), \text{ as } t \to 0,
\end{aligned}$$

and, thus,

$$\frac{\Delta_k(f, x, t)}{t^k} = f_{(k)}(x) + o(1), \text{ as } t \to 0.$$

Hence, $RD_k f(x)$ exists with value $f_{(k)}(x)$.

For the second part, suppose that $\overline{f}_{(k+1)}(x)$ and $\underline{f}_{(k+1)}(x)$ are finite. Then

$$f(x+t) = \sum_{i=0}^{k}\frac{t^i}{i!}f_{(i)}(x) + O(t^{k+1}), \text{ as } t \to 0. \tag{2.22.3}$$

From (2.22.1) and (2.22.3),

$$\begin{aligned}
\Delta_{k+1}(f, x, t) &= \sum_{i=0}^{k+1}(-1)^{k+1-i}\binom{k+1}{i}f(x+it) \\
&= \sum_{i=0}^{k+1}(-1)^{k+1-i}\binom{k+1}{i}\left(\sum_{j=0}^{k}\frac{(it)^j}{j!}f_{(j)}(x) + O(t^{k+1})\right) \\
&= \sum_{j=0}^{k}\frac{t^j}{j!}f_{(j)}(x)\left(\sum_{i=0}^{k+1}(-1)^{k+1-i}\binom{k+1}{i}i^j\right) + O(t^{k+1}) \\
&= O(t^{k+1}), \text{ as } t \to 0.
\end{aligned}$$

Hence, $\overline{RD}_{k+1}f(x)$ and $\underline{RD}_{k+1}kf(x)$ are finite.

For the converse, consider $f(x) = |x|$. Then, since $f_{(1)}(0)$ does not exist, it follows that $f_{(2)}(0)$ does not exist. However, $\lim_{t\to 0\pm}\Delta_2(f,0,t)/t^2 = 0$ and, so, $RD_2f(0)$ exists with value 0. $\qquad\square$

2.23 Symmetric de la Vallée Poussin and Symmetric Riemann Derivatives, $f_{(k)}^{(s)}$ and $RD_k^{(s)}f$

Theorem 2.23.1 *If $f_{(k)}^{(s)}(x)$ exists finitely, then $RD_k^{(s)}f(x)$ exists and is equal to $f_{(k)}^{(s)}(x)$. Further, if $\overline{f}_{(k+2)}^{(s)}(x)$ and $\underline{f}_{(k+2)}^{(s)}(x)$ are finite, then $\overline{RD}_{k+2}^{(s)}f(x)$ and $\underline{RD}_{k+2}^{(s)}f(x)$ also are finite.*

$\square$ We prove the case when k is even, $k = 2m$, say; then, the case of odd k is similar. If $f_{(2m)}^{(s)}(x)$ exists finitely, we have

$$\frac{f(x+t)+f(x-t)}{2} = \sum_{i=0}^{m}\frac{t^{2i}}{(2i)!}f_{(2i)}^{(s)}(x) + o(t^{2m}), \quad \text{as } t \to 0, \qquad (2.23.1)$$

and, so (see (1.2.20) and (1.2.22) of Chapter I), using (2.23.1)

$$
\begin{aligned}
\Delta_{2m}^s(f,x,2t) &= \sum_{i=0}^{2m}(-1)^{2m-i}\binom{2m}{i}f(x+2it-2mt) \\
&= \sum_{i=0}^{2m}(-1)^{2m-i}\binom{2m}{i}f(x-2it+2mt) \\
&= \sum_{i=0}^{2m}(-1)^{2m-i}\binom{2m}{i}\frac{f(x+2it-2mt)+f(x-2it+2mt)}{2} \\
&= \sum_{i=0}^{2m}(-1)^{2m-i}\binom{2m}{i}\left(\sum_{j=0}^{m}\frac{(2it-2mt)^{2j}}{(2j)!}f_{(2j)}^{(s)}(x) + o(t^{2m})\right) \\
&= \sum_{j=0}^{m}2^{2j}\frac{t^{2j}}{(2j)!}f_{(2j)}^{(s)}(x)\left(\sum_{i=0}^{2m}(-1)^{2m-i}\binom{2m}{i}(i-m)^{2j} + o(t^{2m})\right) \\
&= 2^{2m}\frac{t^{2m}}{(2m)!}f_{(2m)}^{(s)}(x)(2m)! + o(t^{2m}), \quad \text{as } t \to 0, \text{ using } (2.22.1).
\end{aligned}
$$

$$(2.23.2)$$

Hence,

$$\frac{\Delta_{2m}^s(f,x,2t)}{(2t)^{2m}} = f_{(2m)}^{(s)}(x) + o(1) \quad \text{as } t \to 0,$$

and, so, $RD_{2m}^{(s)} f(x)$ exists and is equal to $f_{(2m)}^{(s)}(x)$. Next, suppose that $\overline{f}_{(2m+2)}^{(s)}(x)$ and $\underline{f}_{-(2m+2)}^{(s)}(x)$ are finite. Then,

$$\frac{f(x+t) + f(x-t)}{2} = \sum_{i=0}^{m} \frac{t^{2i}}{(2i)!} f_{(2i)}^{(s)}(x) + O(t^{2m+2}), \text{ as } t \to 0. \quad (2.23.3)$$

So, as above, applying (2.23.3) we get,

$$\Delta_{2m+2}^{s}(f, x, 2t) = \sum_{i=0}^{2m+2} (-1)^{2m+2-i} \binom{2m+2}{i} f(x + 2it - 2mt - 2t)$$

$$= \sum_{i=0}^{2m+2} (-1)^{2m+2-i} \binom{2m+2}{i} \frac{f(x+2it-2mt-2t) + f(x-2it+2mt+2t)}{2}$$

$$= \sum_{i=0}^{2m+2} (-1)^{2m+2-i} \binom{2m+2}{i} \left(\sum_{j=0}^{m} \frac{(2i-2m-2)^{2j}}{(2j)!} t^{2j} f_{(2j)}^{(s)}(x) + O(t^{2m+2}) \right)$$

$$= \sum_{j=0}^{m} \frac{(2t)^{2j}}{(2j)!} f_{(2j)}^{(s)}(x) \left(\sum_{i=0}^{2m+2} (-1)^{2m+2-i} \binom{2m+2}{i} (i-m-1)^{2j} + O(t^{2m+2}) \right)$$

$$= O(t^{2m+2}), \text{ as } t \to 0, \text{ using } (2.22.1).$$

So,

$$\frac{\Delta_{2m+2}^{s}(f, x, 2t)}{(2t)^{2m+2}} = O(1), \text{ as } t \to 0.$$

Hence, $\limsup_{t \to 0} \dfrac{\Delta_{2m+2}^{s}(f, x, 2t)}{(2t)^{2m+2}}$ and $\limsup_{t \to 0} \dfrac{\Delta_{2m+2}^{s}(f, x, 2t)}{(2t)^{2m+2}}$ are both finite; that is, $\overline{RD}_{2m+2}^{(s)} f(x)$ and $\underline{RD}_{2m+2}^{(s)} f(x)$ are finite. $\square$

The converse of the above theorem is known only for $k = 1, 2, 3, 4$. From the definition, $\overline{RD}_k^{(s)} f(x) = \overline{f}_{(k)}^{(s)}(x)$ and $\underline{RD}_k^{(s)} f(x) = \underline{f}_{-(k)}^{(s)}(x)$ when $k = 1, 2$. If $k \geq 3$, the definitions of the two derivatives differ; we prove the converse when $k = 3, 4$.

Theorem 2.23.2 *Let $RD_1^s f(x)$ exist finitely. If $RD_3^s f(x)$ exists finitely, then $f_{(3)}^{(s)}(x)$ exists and is equal to $RD_3^s f(x)$. Moreover, if $\underline{RD}_3^s f(x)$ and $\overline{RD}_3^s f(x)$ are finite, then $\underline{f}_{-(3)}^{(s)}(x)$ and $\overline{f}_{(3)}^{(s)}(x)$ are also finite.*

$\square$ Without loss in generality, we may assume that $x = 0$ and first we suppose that

$$RD_1^s f(0) = RD_3^s f(0) = 0. \quad (2.23.4)$$

From (2.23.4), we have

$$f(t) - f(-t) = o(t), \text{ as } t \to 0 \quad (2.23.5)$$

and
$$f(3t) - f(-3t) - 3(f(t) - f(-t)) = o(t^3), \text{ as } t \to 0. \tag{2.23.6}$$

Replacing t by $t/3^{n+1}$ in (2.23.6), we get

$$f(t/3^n) - f(-t/3^n) - 3(f(t/3^{n+1}) - f(-t/3^{n+1})) = o(t^3/3^{3n+3}) \text{ as } t \to 0, \tag{2.23.7}$$

or, since $o(t^3/3^{3n+3})$ as $t \to 0$ implies $3^{-(3n+3)}o(t^3)$ as $t \to 0$,

$$3^n(f(t/3^n) - f(-t/3^n)) - 3^{n+1}(f(t/3^{n+1}) - f(-t/3^{n+1}))$$
$$= \frac{1}{3^{2n+3}}o(t^3) \text{ as } t \to 0. \tag{2.23.8}$$

Putting $n = 0, 1, \ldots, N$ and adding in (2.23.8),

$$(f(t) - f(-t)) - 3^{N+1}(f(t/3^{N+1}) - f(-t/3^{N+1})) = o(t^3) \sum_{n=0}^{N} \frac{1}{3^{2n+3}}. \tag{2.23.9}$$

By (2.23.5), for fixed t

$$\lim_{N \to \infty} 3^{N+1}(f(t/3^{N+1}) - f(-t/3^{N+1})) = 0. \tag{2.23.10}$$

Therefore, letting $N \to \infty$, we have from (2.23.9) that as $t \to 0$

$$\frac{f(t) - f(-t)}{2} = o(t^3)\frac{1}{2}\sum_{n=0}^{\infty} \frac{1}{3^{2n+3}} = o(t^3), \tag{2.23.11}$$

which shows that $f_{(3)}^{(s)}(0)$ exists with value 0.

For the second part let $\underline{RD}_3^s f(0)$ and $\overline{RD}_3^s f(0)$ be finite. Then, instead of (2.23.4), we now consider $RD_1^s f(0) = 0$. Relation (2.23.5) remains true and on the right-hand sides of (2.23.6),(2.23.7), (2.23.8) and (2.23.9) the little-oh terms are replaced by similar big-oh terms; (2.23.10) remains the same as it only uses (2.23.5). So, finally (2.23.11) becomes

$$\frac{f(t) - f(-t)}{2} = O(t^3)\frac{1}{2}\sum_{n=0}^{\infty} \frac{1}{3^{2n+3}} = O(t^3). \tag{2.23.12}$$

It follows from (2.23.12) that $\underline{f}_{(3)}^{(s)}(0)$ and $\overline{f}_{(3)}^{(s)}(0)$ are finite, completing the proof of the special case.

For the general case, we begin by considering the first part of the theorem and define

$$g(x) = f(x) - xRD_1^s f(0) - \frac{x^3}{3!}RD_3^s f(0).$$

Then, g satisfied the conditions of the special case and, so $g_{(3)}^{(s)}(0)$ exists and $g_{(3)}^{(s)}(0) = 0$. So, $f_{(3)}^{(s)}(0)$ exists and $f_{(3)}^{(s)}(0) = RD_3^s f(0)$.

For the second part, put

$$g(x) = f(x) - xRD_1^s f(0).$$

Then, $RD_1^s g(0) = 0$ and using the special case $\overline{g}_{(3)}^{(s)}(0)$ and $\underline{g}_{(3)}^{(s)}(0)$ are finite and, hence, $\overline{f}_{(3)}^{(s)}(0)$ and $\underline{f}_{(3)}^{(s)}(0)$ are finite. □

Theorem 2.23.3 *Let $RD_2^s f(x)$ exists finitely. If $RD_4^s f(x)$ exists finitely, then $f_{(4)}^{(s)}(x)$ exists and is equal to $RD_4^s f(x)$. If $\overline{RD}_4^s f(x)$ and $\underline{RD}_4^s f(x)$ are finite, then $\overline{f}_{(4)}^{(s)}(x)$ and $\underline{f}_{(4)}^{(s)}(x)$ are also finite.*

□   Without loss in generality, we may suppose that $x = 0$ and first consider the special case

$$f(0) = RD_2^s f(0) = RD_4^s f(0) = 0. \tag{2.23.13}$$

From (2.23.13),

$$f(t) + f(-t) = o(t^2), \text{ as } t \to 0, \tag{2.23.14}$$
$$f(2t) + f(-2t) - 4(f(t) + f(-t)) = o(t^4), \text{ as } t \to 0. \tag{2.23.15}$$

In (2.23.15), replace t by $t/2^{n+1}$ to get

$$f(t/2^n) + f(-t/2^n) - 4(f(t/2^{n+1}) + f(-t/2^{n+1})) = o(t^4/2^{4n+4}),$$

and, as in (2.23.8),

$$2^{2n}(f(t/2^n) + f(-t/2^n)) - 2^{2n+2}(f(t/2^{n+1}) + f(-t/2^{n+1}))$$
$$= \frac{1}{2^{2n+4}} o(t^4), \text{ as } t \to 0. \tag{2.23.16}$$

Putting $n = 0, 1, \ldots, N$ in (2.23.16) and adding gives

$$f(t) + f(-t) - 2^{2N+2}(f(t/2^{N+1}) + f(-t/2^{N+1})) = o(t^4) \sum_{n=0}^{N} \frac{1}{2^{2n+4}}. \tag{2.23.17}$$

Now, by (2.23.14), we have for fixed t

$$\lim_{N \to \infty} 2^{2N+2}(f(t/2^{N+1}) + f(-t/2^{N+1})) = 0. \tag{2.23.18}$$

So, letting $N \to \infty$ in (2.23.17), we get from (2.23.18) that

$$\frac{f(t) + f(-t)}{2} = o(t^4) \frac{1}{2} \sum_{n=0}^{\infty} \frac{1}{2^{2n+4}} = o(t^4), \text{ as } t \to 0. \tag{2.23.19}$$

This shows that $f_{(4)}^{(s)}(0)$ exists with value 0.

For the second part of the special case, replace (2.23.13) by $f(0) = RD_2^s f(0) = 0$. Then (2.23.14) is true and the right-hand sides of (2.23.15),(2.23.16), and (2.23.17) the little-oh terms are replaced by similar big-oh terms; (2.23.18) remains the same as it only uses (2.23.14). So, finally (2.23.19) becomes

$$\frac{f(t) + f(-t)}{2} = O(t^4) \frac{1}{2} \sum_{n=0}^{\infty} \frac{1}{2^{2n+4}} = O(t^4), \text{ as } t \to 0. \qquad (2.23.20)$$

From (2.23.20), $\overline{f}_{(4)}^{(s)}(0)$ and $\underline{f}_{(4)}^{(s)}(0)$ are finite, completing the proof of the special case.

For the general case, we begin by considering the first part of the theorem and define

$$g(x) = f(x) - f(0) - \frac{x^2}{2} RD_2^s f(0) - \frac{x^4}{4!} RD_4^s f(0). \qquad (2.23.21)$$

Then, g satisfied the conditions of the special case and, so, $g_{(4)}^{(s)}(0)$ exists and $g_{(4)}^{(s)}(0) = 0$. So, $f_{(4)}^{(s)}(0)$ exists and, from (2.21), $f_{(4)}^{(s)}(0) = RD_4^s f(0)$.

For the second part, put

$$g(x) = f(x) - f(0) - \frac{x^2}{2} RD_2^s f(0). \qquad (2.23.22)$$

Then, g satisfies the conditions of the second part of the special case and using the special case $\overline{g}_{(4)}^{(s)}(0)$ and $\underline{g}_{(4)}^{(s)}(0)$ are finite and, hence, from (2.23.22), $\overline{f}_{(4)}^{(s)}(0)$ and $\underline{f}_{(4)}^{(s)}(0)$ are finite. $\square$

2.24 Generalized Riemann and Peano Derivatives, $GRD_k f$ and $f_{(k)}$

Since the existence of the Peano derivative, $f_{(n)}(x)$, implies the existence of the d.l.V.P. derivative, $f_{(n)}^{(s)}(x)$ (see Theorem 2.5.1), it follows from Theorem 2.23.1 that the existence of $f_{(n)}(x)$ implies the existence of the symmetric Riemann derivative, $RD_n^s f(x)$, and their values will be the same. Also, a particular choice of the system $S = \{a_0, a_1, \ldots, a_{n+\ell}; A_0, A_1, \ldots, A_{n+\ell}; L\}$ in the definition of the generalized Riemann derivative, $GDR_n f(x)$, (see Section 3 of Chapter I), gives the unsymmetric Riemann derivative, $RD_n f(x)$, and the symmetric Riemann derivative, $RD_n^s f(x)$. However the general system S may give a derivative that differs from both $RD_n f(x)$ and $RD_n^s f(x)$, so we prove the following.

Theorem 2.24.1 If $f_{(n)}(x)$ exists finitely, then $GRD_n f(x, S)$ exists and equals $f_{(n)}(x)$ for any system S. Further, if $\overline{f}_{(n)}(x)$ and $\underline{f}_{(n)}(x)$ are finite, then $\overline{GRD}_n f(x, S)$ and $\underline{GRD}_n f(x, S)$ are also finite.

☐   Let $f_{(n)}(x)$ exist finitely. Then

$$f(x+h) = \sum_{i=0}^{n} \frac{h^i}{i!} f_{(i)}(x) + o(h^n), \text{ as } h \to 0. \tag{2.24.1}$$

From (2.24.1) and relation (1.3.1) of Chapter I,

$$
\begin{aligned}
\frac{n!}{Lh^n} \sum_{i=0}^{n+\ell} A_i f(x + a_i h) &= \frac{n!}{Lh^n} \sum_{i=0}^{n+\ell} A_i \left(\sum_{j=0}^{n} \frac{(a_i h)^j}{j!} f_{(j)}(x) + o(h^n) \right) \\
&= \frac{n!}{Lh^n} \sum_{j=0}^{n} \frac{h^j}{j!} f_{(j)}(x) \left(\sum_{i=0}^{n+\ell} A_i a_i^j \right) + o(1) \\
&= \frac{n!}{Lh^n} \frac{h^n}{n!} f_{(n)}(x) L + o(1) = f_{(n)}(x) + o(1), \text{ as } h \to 0.
\end{aligned}
$$

So, letting $h \to 0$

$$GRD_n f(x) = \lim_{h \to 0} \frac{n!}{Lh^n} \sum_{i=0}^{n+\ell} A_i f(x + a_i h) = f_{(n)}(x).$$

For the second part, (2.24.1) is replaced by

$$f(x+h) = \sum_{i=0}^{n-1} \frac{h^i}{i!} f_{(i)}(x) + O(h^n), \text{ as } h \to 0. \tag{2.24.2}$$

From (2.24.2)and relation (1.3.1) of Chapter I,

$$
\begin{aligned}
\frac{n!}{Lh^n} \sum_{i=0}^{n+\ell} A_i f(x + a_i h) &= \frac{n!}{Lh^n} \sum_{i=0}^{n+\ell} A_i \left(\sum_{j=0}^{n-1} \frac{(a_i h)^j}{j!} f_{(j)}(x) + O(h^n) \right) \\
&= \frac{n!}{Lh^n} \sum_{j=0}^{n-1} \frac{h^j}{j!} f_{(j)}(x) \left(\sum_{i=0}^{n+\ell} A_i a_i^j \right) + O(1) \tag{2.24.3} \\
&= O(1), \text{ as } h \to 0.
\end{aligned}
$$

So, letting $h \to 0$, $\overline{GRD}_n f(x, S)$ and $\underline{GRD}_n f(x, S)$, respectively, the upper and lower limits of the left-hand side of (2.24.3) are finite.   ☐

2.25   MZ- and Peano Derivatives, $\tilde{D}_k f$ and $f_{(k)}$

The MZ-derivative of order k at x, $\tilde{D}_k f(x)$, is defined in Section 3 of Chapter I; it is a special generalized Riemann derivative, namely one with

$$S = \{0, 1, 2, 2^2, \ldots, 2^{k-1}; A_0, A_1, \ldots, A_k; L\};$$

$$L = \sum_{i=1}^{k} 2^{(i-1)k} A_i; \quad \ell = 0; \tag{2.25.1}$$

A_i, $0 \le i \le k$, satisfying the equations (1.3.3) of Section 3 of Chapter I.
So,

$$
\begin{aligned}
\tilde{D}_k f(x) &= \lim_{t \to 0} \frac{\lambda_k}{t^k} \tilde{\Delta}_k(f, x, t) \\
&= \lim_{t \to 0} \frac{\lambda_k}{t^k} \left(A_0 f(x) + \sum_{i=1}^{k} A_i f(x + 2^{i-1} t) \right),
\end{aligned}
\tag{2.25.2}
$$

provided the limit exists, where

$$\lambda_k = \frac{k!}{\sum_{i=1}^{k} 2^{(i-1)k} A_i}. \tag{2.25.3}$$

For details of the relevant calculations, see Section 3 of Chapter I.

If the limit does not exist, then the MZ-derivates of f at x of order k are defined to be

$$
\begin{aligned}
\overline{\tilde{D}}_k f(x) &= \limsup_{t \to 0} \frac{\lambda_k}{t^k} \tilde{\Delta}_k(f, x, t), \\
\underline{\tilde{D}}_k f(x) &= \liminf_{t \to 0} \frac{\lambda_k}{t^k} \tilde{\Delta}_k(f, x, t).
\end{aligned}
\tag{2.25.4}
$$

Theorem 2.25.1 If $\tilde{D}_1 f(x), \tilde{D}_2 f(x), \ldots \tilde{D}_k f(x)$ exist finitely, then $f_{(k)}(x)$ exists and $f_{(i)}(x) = \tilde{D}_i f(x)$, $1 = 1, 2, \ldots, k$ and

$$\underline{\tilde{D}}_{k+1} f(x) \le \underline{f}_{(k+1)}(x) \le \overline{f}_{(k+1)}(x) \le \overline{\tilde{D}}_{k+1} f(x). \tag{2.25.5}$$

In particular, if $\tilde{D}_{k+1} f(x)$ exists, possibly infinite, then $f_{(k+1)}(x)$ exists with the same value.

☐   We may suppose without loss in generality that $x = 0$. Let

$$g(t) = f(t) - f(0) - \sum_{i=1}^{k} \frac{t^i}{i!} \tilde{D}_i f(0). \tag{2.25.6}$$

Then, from the definition of the MZ-derivative, it follows that

$$g(0) = \tilde{D}_1 g(0) = \cdots = \tilde{D}_k g(0) = 0. \tag{2.25.7}$$

Let $\epsilon > 0$ be arbitrary. Then, using the last equality in (2.25.7), there is a $\delta > 0$ such that

$$\left|\tilde{\Delta}_k(g,0,t)\right| < \epsilon|t|^k, \quad \text{for } |t| < \delta. \tag{2.25.8}$$

Since $\tilde{\Delta}_k(g,0,t) = \tilde{\Delta}_{k-1}(g,0,2t) - 2^{k-1}\tilde{\Delta}_{k-1}(g,0,t)$ (see relation (1.3.7) of Chapter I), we have from (2.25.8)

$$\left|\tilde{\Delta}_{k-1}(g,0,2t) - 2^{k-1}\tilde{\Delta}_{k-1}(g,0,t)\right| < \epsilon|t|^k, \quad \text{for } |t| < \delta. \tag{2.25.9}$$

Keeping t fixed in (2.25.9) and replacing t by $t/2^{n+1}$, we get

$$\left|\tilde{\Delta}_{k-1}(g,0,t/2^n) - 2^{k-1}\tilde{\Delta}_{k-1}(g,0,t/2^{n+1})\right| < \frac{\epsilon|t|^k}{2^{k(n+1)}}, \quad \text{for } |t| < \delta, \tag{2.25.10}$$

or on multiplying (2.25.10) by $2^{(k-1)n}$

$$\left|2^{(k-1)n}\tilde{\Delta}_{k-1}(g,0,t/2^n) - 2^{(k-1)(n+1)}\tilde{\Delta}_{k-1}(g,0,t/2^{n+1})\right| < \frac{\epsilon|t|^k}{2^{k+n}}. \tag{2.25.11}$$

Putting $n = 0,1,\ldots,N$ in (2.25.11) and adding gives

$$\left|\tilde{\Delta}_{k-1}(g,0,t) - 2^{(k-1)(N+1)}\tilde{\Delta}_{k-1}(g,0,t/2^{N+1})\right| < \frac{\epsilon|t|^k}{2^k} \sum_{n=0}^{N} \frac{1}{2^n}. \tag{2.25.12}$$

Since, from (2.25.7), $\tilde{D}_{k-1}g(0) = 0$, we have

$$\lim_{N\to\infty} 2^{(k-1)(N+1)}\tilde{\Delta}_{k-1}(g,0,t/2^{N+1}) = 0. \tag{2.25.13}$$

So, letting $N \to \infty$ in (2.25.12), we have from (2.25.13) that

$$\left|\tilde{\Delta}_{k-1}(g,0,t)\right| \le \frac{\epsilon|t|^k}{2^{k-1}} < \epsilon|t|^k, \quad \text{for } |t| < \delta. \tag{2.25.14}$$

We have deduced (2.25.14) from (2.25.8) using (2.25.7) and the relation (1.3.7) of Chapter I, quoted above. In a similar, way we can deduce the following from (2.25.14):

$$\left|\tilde{\Delta}_{k-2}(g,0,t)\right| < \epsilon|t|^k, \quad \text{for } |t| < \delta,$$

and, ultimately, after a few steps we get

$$\left|\tilde{\Delta}_1(g,0,t)\right| < \epsilon|t|^k, \quad \text{for } |t| < \delta. \tag{2.25.15}$$

Since, from (2.25.7), $g(0) = 0$ (2.25.15) gives

$$\left|g(t)\right| < \epsilon|t|^k, \quad \text{for } |t| < \delta. \tag{2.25.16}$$

Hence, $g_{(k)}(0)$ exists and $g_{(i)}(0) = 0$, $i = 1, 2, \ldots, k$. Thus, by (2.25.6), $f_{(k)}(0)$ exists and $f_{(i)}(0) = \widetilde{D}_i f(0)$, $i = 1, 2, \ldots, k$. This completes the proof of the first part.

For the second part, again let $x = 0$, and prove the last inequality as the proof of the first is similar. If $\overline{\overline{D}}_{k+1} f(0) = \infty$, there is nothing to prove, so we suppose that $\overline{\overline{D}}_{k+1} f(0) = M < \infty$. Consider

$$g(t) = f(t) - f(0) - \sum_{i=1}^{k} \frac{t^i}{i!} \widetilde{D}_i f(0) - \frac{t^{k+1}}{(k+1)!} M. \qquad (2.25.17)$$

Since $f_{(i)}(0)$ exists, $i = 1, 2, \ldots, k$, we get from (2.25.17) that $g_{(i)}(0)$ exists, $g_{(i)}(0) = 0$ $i = 1, 2, \ldots, k$, and $\overline{\overline{D}}_{k+1} g(0) = \overline{\overline{D}}_{k+1} f(0) - M = 0$. Let $\epsilon > 0$ be arbitrary, then there exists a $\delta > 0$ such that

$$\widetilde{\Delta}_{k+1}(g, 0, t) < \epsilon t^{k+1}, \quad \text{for } 0 < t < \delta. \qquad (2.25.18)$$

The relation (2.25.18) is analogous to (2.25.8) and, so, applying arguments similar to those that deduced (2.25.16) from (2.25.8), we deduce from (2.25.18) that

$$g(t) < \epsilon t^{k+1}, \quad \text{for } 0 < t < \delta. \qquad (2.25.19)$$

Hence, $\overline{g}^+_{(k+1)}(0) \leq (k+1)!\epsilon$, and since ϵ is arbitrary, this gives $\overline{g}^+_{(k+1)}(0) \leq 0$ or, equivalently, $\overline{f}^+_{(k+1)}(0) \leq M$.

For the negative side, note that when $k+1$ is odd, the relation (2.25.18) becomes

$$\widetilde{\Delta}_{k+1}(g, 0, t) > \epsilon t^{k+1}, \quad \text{for } -\delta < t < 0.$$

From this, we get $g(t) > \epsilon t^{k+1}$, for $-\delta < t < 0$, which implies $\overline{g}^-_{(k+1)}(0) \leq 0$ or $\overline{f}^-_{(k+1)}(0) \leq M$.

Thus, $\overline{f}_{(k+1)}(0) \leq M$, which completes the proof. $\qquad \square$

Remark. The existence of $\widetilde{D}_k f(x)$ does not imply the existence of $\widetilde{D}_{k-1} f(x)$. To see this, consider

$$f(x) = \begin{cases} x, & \text{if } x \geq 0, \\ 0, & \text{if } x < 0. \end{cases}$$

Here, $\widetilde{D}_1 f(0)$ does not exists, but $\widetilde{D}_2 f(0)$ does, $\widetilde{D}_2 f(0) = 0$.

Theorem 2.25.2 *If* $f_{(k)}(x)$ *exists finitely, then* $\widetilde{D}_k f(x)$ *exists with value* $f_{(k)}(x)$. *Moreover, if* $\overline{f}_{(k+1)}(x)$ *and* $\underline{f}_{(k+1)}(x)$ *are finite, then* $\overline{\overline{D}}_{k+1} f(x)$ *and* $\underline{\underline{D}}_{k+1} f(x)$ *are also finite.*

$\square$ $\quad$ Since the MZ- derivatives and derivates are special cases of the generalized Riemann derivatives and derivates considered in the previous section, this result follows from Theorem 2.24.1. The argument used in that proof can be given for this special case, but the argument is not really any simpler. $\quad \square$

Bibliography

1. S. Agronsky, A. M. Bruckner, M. Laczkovich & D. Preiss, *Convexity conditions and intersections with smooth functions*, Trans. Am. Math. Soc. **289** (1985), 659–677.

2. J. Marshall Ash, *Generalizations of the Riemann derivative*, Trans. Am. Math. Soc. **126** (1967), 181–199.

3. ———, *A characterization of the Peano derivative*, Trans. Am. Math. Soc. **149** (1970), 489–501.

4. ———, *Very generalized Riemann derivatives, generalized Riemann derivatives and associated summability methods*, Real Anal. Exchange **11** (1985/86), 10–29.

5. ———, *An L^p-differentiable non-differentiable function*, Real Anal. Exchange **30** (2004/05), 747–754.

6. ——— & R. L. Jones, *Mean value theorems for generalized Riemann derivatives*, Proc. Am. Math. Soc. **101** (1987), 263–271.

7. H. Auerbach, *Sur les dérivées généralisées* Fund. Math. **8** (1926), 49–55.

8. B. S. Babcock, *On properties of the approximate Peano derivatives*, Trans. Am. Math. Soc. **212** (1975), 279–294.

9. C. L. Belna, M. J. Evans & P. D. Humke, *Symmetric and ordinary differentiation*, Proc. Am. Math. Soc. **72** (1978), 261–267.

10. J. A. Bergin, *A new characterization of Cesàro-Perron integrals using the Peano derivates*, Trans. Am. Math. Soc. **228** (1977), 287–305.

11. G. Brown, *Continuous functions of bounded nth variation*, Proc. Edinburgh Math. Soc. **3** (1968/69), 205–314.

12. A. M. Bruckner, *Differentiation of Real Functions*, Lecture Notes in Mathematics, 659, Springer, Berlin, 1978, 2nd ed., CRM Monograph Series, 5, American Mathematical Society, Providence, RI, 1994.

13. ———, R. J. O'Malley & B. S. Thomson, *Path derivatives: A unified view of certain generalized derivatives*, Trans. Am. Math. Soc. **283** (1984), 97–125.

14. Z. Buczolich, *An extension theorem for higher Peano derivatives in* $\mathbb{R}^n$, Real Anal. Exchange **13** (1987/88), 245–252.

15. ———, *Second Peano derivatives are not extendable*, Real Anal. Exchange **14** (1988/89), 423–428.

16. ———, *Convexity and symmetric derivates of measurable functions*, Real Anal. Exchange **16** (1990/91), 187–196.

17. ———, *Infinite Peano derivatives, extensions and Baire one property*, Atti. Semin. Mat. Fis. Univ. Modena Reggio Emilia **52** (2004), 117–149.

18. ———, M. J. Evans & P. D. Humke, *Approximate higher order smoothness*, Acta Math. Hungar. **61** (1993), 369–388.

19. ——— & C .E. Weil, *The non-coincidence of ordinary and Peano derivatives*, Math. Bohem. **124** (1999), 381–399.

20. P. S. Bullen, *Construction of primitives of generalised derivatives with applications to trigonometric series*, Can. J. Math. **13** (1961), 48–58.

21. ———, *A criterion for n-convexity*, Pac. J. Math. **36** (1971), 81–98.

22. ——— & C. M. Lee, *The* SC_nP-*integral and the* P^{n+1}-*integral*, Can. J. Math. **25** (1973), 1274–1284.

23. ——— & S. N. Muhkopadhyay, *Peano derivatives and general integrals*, Pac. J. Math. **47** (1973), 43–58.

24. ———, *On the Peano derivatives*, Can. J. Math. **25** (1973), 127–140.

25. ———, *Integration by parts formula for some trigonometric integrals*, Proc. London Math. Soc. **(3) 29** (1974), 159–173.

26. ———, *Relations between some general n-th order derivatives*, Fund. Math. **85** (1974), 257–276.

27. ———, *The Peano derivative and the* M_2-*property of Zahorski*, Indian J. Math. **28** (1986), 219–228

28. J. C. Burkill, *The Cesàro-Perron integral*, Proc. London Math. Soc. **(2) 34** (1932), 314–322.

29. ———, *The Cesàro-Perron scale of integration*, Proc. London Math. Soc. **(2) 39** (1935), 541–552.

30. ———, *Integrals and trigonometric series*, Proc. London Math. Soc. **(3) 1** (1951), 46–57.

31. P. L. Butzer & W. Kozakiewicz, *On the Riemann derivatives of integrable functions*, Can. J. Math. **6** (1954), 572–581.

32. A. Calderón & A. Zygmund, *Local properties of solutions of elliptical partial differential equations*, Studia Math. **20** (1961), 171–225.

33. P. S. Chakraborty, *On properties of Peano derivatives*, Acta Math. Acad. Sci. Hung. **32** (1978), 217–228.

34. Z. Charzynski, *Sur les fonctions dont les dérivées symmetrique est partout finie*, Fund. Math. **21** (1931), 214–225.

35. Z. Ciesielski, *Some properties of convex functions of higher orders*, Ann. Polon. Math. **221** (1959), 1–7.

36. E. Corominas, *Dérivation de Riemann-Schwarz*, C. R. Acad. Sci. Paris. Sér. I Math. **224** (1947), 176–177.

37. ———, *Contribution à la théorie de la dérivation d'ordre supérieur* Bull. Soc. Math. France **81** (1953), 177–222.

38. G. E. Cross, *The $SC_{k+1}P$ integral and trigonometric series*, Proc. Am. Math. Soc. **69** (1978), 297–302.

39. ———, *Some condition for n-convex functions*, Proc. Am. Math. Soc. **82** (1981), 587–592.

40. ———, *Continuous linear transformations of n-convex functions*, Real Anal. Exchange **11** (1985/86), 416–426.

41. ———, *The integration of exact Peano derivatives*, Can. Math. Bull. **29** (1986), 334–340.

42. A. Császár, *Sur une généralisation de la notion de derivée*, Acta Sci. Math. Szeged **16** (1955), 137–159.

43. A. G. Das & B .K. Lahiri, *On absolutely k-th continuous functions*, Fund. Math. **105** (1980), 159–169.

44. A. Denjoy, *Sur une propriété des fonctions derivées*, Enseignement Math. **18** (1916), 320–328.

45. ———, *Sur l'intégration des coefficients différentiels d'ordre supérieur*, Fund. Math. **25** (1935), 273–326.

46. T. K. Dutta, *Generalized smooth functions*, Acta Math. Acad. Sci. Hung. **40** (1982), 29–37.

47. ———, *On uniform generalized symmetric derivates*, Indian J. Math. **33** (1991), 63–74.

48. ——— & S. N. Mukhopadhyay, *On the Riemann derivatives of C_sP-integrable functions*, Anal. Math. **15** (1987), 159–174.

49. ———, *Generalized smooth functions II*, Acta Math. Hung. **55** (1990), 47–56.

50. M. J. Evans, L_p-*derivatives and approximate Peano derivatives*, Trans. Am. Math. Soc., **165** (1972), 381–388.

51. ———, *Higher order smoothness*, Acta Math. Hung. **50** (1987), 17–20.

52. ———, *Peano differentiation and high order smoothness in L_p*, Bull. Inst. Math. Acad. Sinica **13** (1985), 197–209.

53. ———, *Approximate Peano derivative and Baire* one property*, Real Anal. Exchange **11** (1985-86), 283–289.

54. ——— & P. D. Humke, *A pathological approximately smooth function*, Acta Math. Hung. **46** (1985), 211–215.

55. ——— & L. M. Larson, *The continuity of symmetric and smooth functions*, Acta Math. Hung. **43** (1984), 251–257.

56. ———, *Monotonicity, symmetry and smoothness*, Classical real analysis (Madison, WI, 1982), Contemp. Math. **42** (1985), 49–54, Am. Math. Soc., Providence, RI.

57. ——— & C. E. Weil, *Peano derivatives*, Real Anal. Exchange **7** (1981/82), 5–23.

58. ———, *On iterated L_p-derivative*, Bull. Inst. Math. Acad. Sinica **10** (1982), 89–94.

59. H. Fejzic, *On generalized Peano and Peano derivatives*, Fund. Math. **143** (1993), 55–74.

60. ———, *Decomposition of Peano derivative*, Proc. Am. Math. Soc. **119** (1993), 599–609.

61. ———, *On approximate Peano derivatives*, Acta Math. Hung. **65** (1994), 319–332.

62. ———, *Infinite approximate Peano derivatives*, Proc. Am. Math. Soc. **131** (2002), 2527–2536.

63. ———, J. Mařík & C. E. Weil, *Extending Peano derivative*, Math. Bohem. **119** (1994), 378–406.

64. ———, C. Freiling & D. Rinne, *A mean value theorem for generalized Riemann derivatives*, Proc. Am. Math. Soc. (electronically published Nov. 6, 2007), S0002-9939(07).

65. A. Fischer, *Differentiability of Peano derivatives*, Proc. Am. Math. Soc. **136** (2008), 1779–1785.

66. K. M. Garg, *A new notion of derivative*, Real Anal. Exchange **7** (1981/82), 65–84.

67. ———, *Theory of Differentiation*, Canadian Mathematical Society Series of Monographs and Advanced Texts, 24, A Wiley-Interscience Publication, John Wiley & Sons, New York, 1998

68. A. Genocchi, *Differentialrechnung und Grundzuge der Integralrechnung herausgegeben von Giuseppe Peano*, 1899; transl. from Italian by G. Bohlman and A. Schepp; mit einem Vorwort von A. Meyer, Leipzig, B. G. Teubner.

69. R. Ger, *Convex functions of higher order in Euclidean spaces*, Ann. Polon. Math. **25** (1971/72), 293–302.

70. I. Ginchev, A. Guerraggio & M. Rocca, *Equivalence of (n+1)th order Peano and usual derivative for n-convex functions*, Real Anal. Exchange **25** (1999/00), 513–520.

71. ——— & M. Rocca, *On Peano and Riemann derivatives*, Rend. Circ. Mat. Palermo **49** (2000), 463–480.

72. N. Giovannelli, *On a generalization of the k-pseudosymmetric derivative*, Accad. Sci. Lett. Arti. Palermo **ser(5), 38** (1978), 219–229.

73. E. Görlich & R. Nessel, *Über Peano- und Riemann-Ableitungen in der Norm*, Arch. Math.(Basel) **18** (1967), 399–410.

74. A. Guerraggio & M. Rocca, *Derivate dirette di Riemann e di Peano, Convessita e Calcolo Parallelo (Eds. G. Giogi and F. A. Rossi)*, (Verona, 1997).

75. P. D. Humke & M. Laczkovitch, *Monotonicity theorems for generalized Riemann derivatives*, Rend. Circ. Mat. Palermo **38** (1989), 437–454.

76. ———, *Convexity theorems for generalized Riemann derivatives*, Real Anal. Exchange **15** (1989/90), 652–674.

77. ——— & T. Šalát, *Remarks on strong and symmetric differentiabity of real functions*, Acta Math. Univ. Comenian **51/52** (1987), 235–241.

78. R. D. James, *Generalized nth primitive*, Trans. Am. Math. Soc. **76** (1954), 149–176.

79. J. M. Jedrzewski, *Approximately smooth functions*, Zeszyty Nauk. Univ. Łódz. Nauk. Mat. Przyrod. Ser. II 52 (1973), 7–14.

80. I. P. Kärkliya, *On quasi smooth functions*, Latvijas Valsts Univ. Zinätn. Ratski **41** (1961), 43–46.

81. C. Kassimatis, *Functions which have generalized Riemann derivatives*, Can. J. Math. **10** (1958), 413-420.

82. ———, *On the de la Vallée Poussin derivative*, Proc. Am. Math. Soc. **16** (1965), 1171–1172.

83. ———, *On generalized derivatives*, Studia Math. **25** (1965), 369–371.

84. ———, *Generalized derivatives and de la Vallée Poussin derivative*, Proc. Am. Math. Soc **17** (1966), 633–655.

85. J. H. B. Kemperman, *On the regularity of generalized convex functions*, Trans. Am. Math. Soc. **135** (1969), 69–93.

86. A. Khintchine, *Recherches sur la structure des fonctions mesurables*, Fund. Math. **9** (1927), 212–279.

87. P. Kostyrko, *On the symmetric derivative*, Colloq. Math. **25** (1972), 265–267.

88. N. K. Kundu, *On uniform symmetric derivatives*, Colloq. Math **24** (1972), 259–269.

89. ———, *Symmetric derivatives and approximate symmetric derivatives of continuous functions*, Rev. Roumaine Math. Pures et Appl. **18** (1973), 689–697.

90. ———, *Denjoy theorem for symmetric derivatives*, Rev. Roumaine Math. Pures et Appl. **18** (1973), 699–700.

91. ———, *On some properties of symmetric derivatives*, Ann. Polon. Math. **30** (1974), 9–18.

92. ———, *On symmetric derivatives and on the properties of Zahorski*, Czech. Math. J. **26** (1976), 154–160.

93. M. Lazcovitch, *On the absolute Peano derivatives*, Ann. Univ. Sci. Budapest Eötvös, Sect. Math. **21** (1978), 83–97.

94. ———, *Infinite Peano derivative*, Real Anal. Exchange **26** (2000/01), 811–825.

95. ———, D. Preiss & C. E. Weil, *On unilateral and bilateral nth Peano derivatives*, Proc. Am. Math. Soc. **99** (1987), 129–134.

96. L. Larson, *A method for showing that general derivatives are in Baire class 1*, Classical real analysis (Madison, WI, 1982), Contemp. Math. 42, Am. Math. Soc., Providence, RI, **42** (1985), 87–95.

97. C. M. Lee, *On functions with summable approximate Peano derivatives*, Proc. Am. Math. Soc. **57** (1976), 53–57.

98. ———, *On approximate Peano derivatives*, J. London Math. Soc. **(2)** **12** (1976), 475–478.

99. ———, *Generalization of Cesàro continuous functions and integrals of Perron type*, Trans. Am. Math. Soc. **266** (1981), 461–481.

100. ———, *On absolute Peano derivatives*, Real Anal. Exchange **8** (1982/83), 228–243.

101. ———, *On generalized Peano derivatives*, Trans. Am. Math. Soc. **275** (1983), 381–396.

102. ———, *On generalizations of exact Peano derivative*, Classical real analysis (Madison, WI, 1982) 97–103, Contemp. Math. 42, Am. Math. Soc., Providence RI, **43** (1985), 97–103.

103. ———, & R. J. O'Malley, *The second approximate Peano derivative and the second approximate derivative*, Bull. Inst. Math. Acad. Sinica **3** (1975), 193–197.

104. N. C. Manna, *On a generalization of smooth and symmetric functions*, Colloq. Math. **22** (1971), 291–299.

105. J. Marcinkiewicz, *Sur les séries de Fourier*, Fund. Math. **27** (1937), 38–69.

106. ———, & A. Zygmund, *On the differentiability of functions and summability of trigonometric series*, Fund. Math. **26** (1936), 1–43.

107. J. Mařík, *On generalized derivatives*, Real Anal. Exchange **3** (1977/78), 87–92.

108. S. Mazurkiewicz, *On the first generalized derivative* (in Polish), Prace Mat.-Fiz. **28** (1917), 79–85.

109. ———, *On the relation between the existence of the second generalized derivative and the continuity of a function* (in Polish), Prace Mat.-Fiz. **30** (1919), 225–242.

110. ———, *Sur la dérivée prémière géneralisée*, Fund. Math. **11** (1928), 145–147.

111. M. R. Mewhdi, *On convex functions*, J. London Math. Soc. **39** (1964), 321–326.

112. S. Mitra & S. N. Mukhopadhyay, *Convexity conditions for generalized Riemann derivable functions*, Acta Math. Hung. **83(4)** (1999), 267–191.

113. ———, *Derivates, approximate derivates and porosity derivates of n-convex functions*, Real Anal. Exchange **27** (2001/02), 249–259.

114. S. K. Mukhopadhyay & S. N. Mukhopadhyay, *Functions of bounded kth variation and absolutely kth continuous functions*, Bull. Aust. Math. Soc. **46** (1992), 91–106.

115. S. N. Mukhopadhyay, *On Schwarz differentiablity II*, Rev. Roumaine Math. Pures et Appl. **9** (1964), 859–862.

116. ———, *On Schwarz differentiablity IV*, Acad. Math. Acad. Sci. Hung. **17** (1966), 129–136.

117. ———, *On a certain property of derivatives*, Fund. Math. **67** (1970), 279–284.

118. ———, *On a property of approximate derivatives*, Acta Sci. Szeged **33** (1972), 207–210.

119. ———, *On the regularity of the P^n-integral and its application to trigonometric series*, Pac. J. Math. **55** (1974), 233–247.

120. ———, *On the approximate Peano derivatives*, Fund. Math. **88** (1975), 133–143.

121. ———, *On the Abel summability of trigonometric series*, J. London Math. Soc. **(2) 17** (1978), 87–96.

122. ———, *On the Abel limit of the terms of trigonometric series*, J. London Math. Soc. **(2) 20** (1979), 319–326.

123. ———, & B. K. Lahiri, *On the Schwarz differentiability of continuous functions*, Rev. Roumaine Math. Pures et Appl. **12** (1967), 861–864.

124. ———, & S. Mitra, *Measurability of Peano derivates and approximate Peano derivates*, Real Anal. Exchange **20** (1994/95), 768–775.

125. ———, *An extension of a theorem of Marcinkiewicz and Zygmund on differentiability*, Fund. Math. **151** (1996), 21–38.

126. ———, *An extension of a theorem of Ash on generalized differentiability*, Real Anal. Exchange **24** (1998/99), 351–371.

127. ———, & S. Ray, *On extending Peano derivatives*, Acta Math. Hung. **89** (2000), 327–346.

128. ———, *Baire*-1 and Baire 1 and Zahorski properties of higher order derivatives*, Acta Math. Hung. **112(4)** (2006), 285–305.

129. ———, *Convexity conditions for approximate d.l.V.P. derivable functions*, Indian J. Math. **49** (2007), 71–92.

130. ———, & D. Sain, *On functions of bounded nth variation*, Fund. Math. **131** (1988), 181–208.

131. C. J. Neugebauer, *Smoothness and differentiability in L_p*, Studia Math. **25** (1964), 81–91.

132. ———, *Symmetric, continuous and smooth functions*, Duke Math. J. **31** (1964), 23–32.

133. ———, *Some observations on harmonic, Borel, approximate and L_p differentiability*, Ind. Univ. Math. J. **22** (1972), 5–11.

134. H. W. Oliver, *The exact Peano derivative*, Trans. Am. Math. Soc. **76** (1954), 444–456.

135. R. J. O'Malley & C. E. Weil, *The oscillatory behavior of certain derivatives*, Trans. Am. Math. Soc. **234** (1977), 967–981.

136. ———, *Iterated L_p-derivative*, Bull. Inst. Math. Acad. Sinica **6** (1978), 93–99.

137. B. K. Pal & S. N. Mukhopadhyay, *The Cesàro–Denjoy–Bochner scale of integration*, Acta Math. Hungar. **42** (1983), 243–255.

138. ———, *The Cesàro-Denjoy-Pettis scale of integration*, Acta Math. Hung. **45** (1985), 289–295.

139. G. Peano, *Sulla formola di Taylor*, Atti Accad. Sci. Torino **27** (1891), 40–46.

140. T. Popoviciu, *Sur l'approximation des fonctions convexes d'ordre supérieur*, Mathematica (Cluj) **10** (1935), 49–54.

141. ———, *Les Fonctions Convexes*, Herman et Cie., Paris, 1944.

142. H. W. Pu & H. H. Pu, *On iterated Peano derivatives*, Bull. Inst. Math. Acad. Sinica **7** (1979), 323–328.

143. A. W. Robert & D. E. Varberg, *Convex Functions*, Academic Press, New York, 1973.

144. L. A. Rubel, *A pathological Lebesgue measurable function*, J. London Math. Soc. **38** (1963), 1–4.

145. A. M. Russel, *Functions of bounded kth variation*, Proc. London Math. Soc. **(3) 26** (1973), 547–563.

146. G. Russo & S. Valenti, *On k-pseudosymmetrical approximate differentiability*, Fund. Math. **114** (1981), 79–83.

147. I. Z. Ruszo, *Locally symmetric functions*, Real Anal. Exchange **4** (1978/79), 84–86.

148. S. Saks, *On the generalized derivative J. London Math. Soc. **7** (1932), 247–251.

149. W. L. C. Sargent, *The Borel derivates of a function*, Proc. London Math. Soc. **(2) 38** (1935), 180–196.

150. ———, *On Cesàro derivates of a function*, Proc. London Math. Soc. **(2) 40** (1936), 235–254.

151. ———, *On generalized derivatives and Cesàro–Denjoy integrals*, Proc. London Math. Soc. **(2) 52** (1951), 365–376.

152. ———, *Some property of C_λ-continuous functions*, J. London Math. Soc. **26** (1951), 116–121.

153. H. S. Shapiro, *Monotone singular functions of high smoothness*, Mich. Math. J. **15** (1968), 265–275.

154. W. Sierpinski, *Sur les fonctions convexes mesurable*, Fund. Math. **1** (1920), 125–128.

155. ———, *Sur une hypothèse de Mazurkiewicz* Fund. Math. **11** (1928), 148–150.

156. V. A. Starcev, *The smoothness of functions with respect to a set*, Mat. Zametki **15** (1974), 431–436.

157. E. M. Stein, *Singular Integrals and Differentiability Properties of Functions*, Princeton University Press, Princeton, NJ, 1970.

158. ——— & A. Zygmund, *Smoothness and differentiability of functions*, Ann. Univ. Sci. Budapest Sect. Math. III–IV, (1960–61), 295–307.

159. ———, *On the differentiability of functions*, Studia Math. **23** (1964), 247–283.

160. R. E. Svetic, *The Laplace derivative*, Comment. Math. Univ. Carolina **42** (2001), 331–343.

161. ——— & H. Volkmer, *On the ultimate Peano derivative*, J. Math. Anal. Appl. **218** (1998), 439–452 .

162. E. Szpilrajn, *Remarques sur la dérivée symétrique* Fund Math. **21** (1933), 226–228.

163. S. J. Taylor, *An integral of Perron's type defined with the help of trigonometric series*, Q. J. Math. Oxford **(2) 6** (1955), 255–274.

164. B. S. Thomson, *Some symmetric covering lemmas*, Real Anal. Exchange **15** (1989/90), 346–383.

165. ———, *An analogue of Charzynski's theorem*, Real Anal. Exchange **15** (1989/90), 743–753.

166. ——, *Symmetric variation*, Real Anal. Exchange **17** (1991/92), 409–415.

167. ——, *The range of a symmetric derivative*, Real Anal. Exchange **18** (1992/93), 615–618.

168. ——, *Symmetric Properties of Real Functions*, Marcel Dekker, Inc., New York 1994.

169. A. F. Timan, *Quasi-smooth functions*, Uspehi Matem. Nauk. (N.S.) **5** (1950), 128–130.

170. ——, *On quasi-smooth functions*, Izvestiya Akad. Nauk. SSSR Ser. Mat. **15** (1951), 243–254.

171. D. L. Tore & M. Rocca, *Higher order uniform smoothness and differentiability of real functions*, Real Anal. Exchange **26** (2001/02), 657–668.

172. J. Uher, *Symmetrically differentiable functions are differentiable almost everywhere*, Real Anal. Exchange **8** (1982/83), 253–260.

173. S. Valenti, *Sur la dérivation k-pseudo-symmétrique des fonctions numérique*, Fund. Math. **74** (1972), 147–152.

174. Ch. J. de la Vallée Poussin, *Sur l'approximation des fonction d'une variable réele et de leurs dérivées par les polynomes et les suite limitées de Fourier*, Bull. de l'Acad. Royale Bélgique, (1908), 193–254.

175. S. Verblunsky, *Generalized third derivative and its application to the theory of trigonometric series*, Proc. London Math. Soc. **(2) 31** (1930), 387–406.

176. ——, *Generalized fourth derivative*, J. London Math. Soc. **6** (1931), 82–84.

177. ——, *On the theory of trigonometric series (I)*, Proc. London Math. Soc. **(2) 34** (1932), 441–456.

178. ——, *On the theory of trigonometric series (II)*, Proc. London Math. Soc. **(2) 34** (1932), 457–491.

179. ——, *On the theory of trigonometric series (V)*, Fund. Math. **81** (1933), 168–220.

180. ——, *On the theory of trigonometric series (VI)*, Proc. London Math. Soc. **(2) 38** (1935), 284–326.

181. ——, *On the Peano derivatives*, Proc. London Math. Soc. **(3) 33** (1971), 313–324.

182. ———, *An extension of the inequality of Rajchman–Zygmund*, J. London Math. Soc. **(2) 22** (1980), 87–98.

183. H. Volkmer, *Extending Peano derivative: Necessary and sufficient condition*, Fund. Math. **158** (1999), 219–229.

184. C. Weil, *On properties of derivatives*, Trans. Am. Math. Soc. **114** (1965), 363–376.

185. ———, *On approximate and Peano derivatives*, Proc. Am. Math. Soc. **20** (1969), 487–490.

186. ———, *A property of certain derivatives*, Ind. Univ. Math. J. **23** (1973), 527–536.

187. ———, *A topological lemma and its application to real functions*, Pac. J. Math. **44** (1973), 757–765.

188. ———, *Monotonicity, convexity and symmetric derivatives*, Trans. Am. Math. Soc. **222** (1976), 225–237.

189. ———, *Iterated L_p-derivative*, Real Anal. Exchange **4** (1978/79), 49–51.

190. M. Weiss, *On symmetric derivatives in L_p*, Studia Math. **24** (1964), 89–100.

191. ——— & A. Zygmund, *A note on smooth functions*, Nederl. Akad. Wetensch. Proc. Ser. A **62** (1959) 52–58; Indag. Math. 21 (1959) 52–58.

192. W. Wilczyński, *Some kind of generalized convexity of functions*, Bull. Acad. Polon. Sci. Sect. Math. Astron. et Phys. **20** (1972), 721–723.

193. F. Wolf, *On summable trigonometric series: An extension of uniqueness theorem*, Proc. London Math. Soc. **(2) 45** (1939), 328–356.

194. A. Zygmund, *A theorem on generalized derivatives*, Bull. Am. Math. Soc. **49** (1943), 917–923.

195. ———, *Smooth functions*, Duke Math. J. **12** (1945), 47–76.

196. ———, *Trigonometric Series Vols. I, II*, Cambridge University Press, Cambridge, U.K., 1968.

197. H. Fejzic, R. E. Svetic & C. E. Weil, *Differentiation of n-convex functions*, Fund. Math. **209** (2010), 9–25.

198. S. N. Mukhhopadhyay & S. Ray, *On Laplace derivative*, Anal. Math. **36** (2010), 131–153.

Index

Milton Keynes UK
Ingram Content Group UK Ltd.
UKHW040059071024
449327UK00019B/669